Handbook of Pipes and Piping Bursts, Stress and Strains, Analysis and Control

Handbook of Pipes and Piping Bursts, Stress and Strains, Analysis and Control

Gundel Wilson

Oxford Book Company
Jaipur ▪ New Delhi

Published By

Oxford Book Company

Regd. Off. "Bony Residency", Gate No. 2
Opp. Tilak Public School, Vishweswariya Nagar,
Gopalpura Road, Jaipur - 302018, India
Ph: 0141 - 2761280, Telefax: 2761381
email - oxfordbookjpr@gmail.com
website - oxfordbookcompany.com

Delhi Office: 102, 1st Floor, Satyam House,
4327/3, Ansari Road, Darya Ganj,
New Delhi - 110002, India
Ph: 011-45652440

Distributed by
KEYA INTERNATIONAL PRESS LLP
E-mail: keyainternationalpressllp@gmail.com

Published in India, 2026

ISBN : 9789355241030

Typeset by Shivangi Computers, Jaipur

Printed and bound in India by Replika Press Pvt. Ltd.

Contents

Preface

Mechanical damage to pipelines may take the form of dents and gouges, which may also incorporate cracks. The combination of a gouge within a dent has long been recognized as the most severe form of damage, and can result in failure at very low internal pressures. Yet despite much expensive full-scale testing, predictions of subsequent burst pressures of such damaged pipes are unreliable, and assessment procedures are very conservative. The universal viscoelastic model was further validated in this study using two acrylonitrile–butadiene–styrene (ABS) viscoelastic materials to better elucidate the direct relationship between the failure criterion characteristics involving creep, constant strain rate, stress relaxation, and pipe burst. Using the yield strain as the failure criterion for constant strain rate and stress relaxation measurements and the strain at critical creep, the failure condition for creep, it was found that the universal viscoelastic model allowed these failure criteria to yield remarkably good agreement on a projected time scale. The relationship of the failure criteria between these three different techniques for characterizing a viscoelastic material was successfully used to identify several complementary approaches to predict long-term pipe burst for two different ABS materials.

The pipe burst data for ABS-A appeared to fit the Tresca failure criterion better than the von Mises failure criterion, as predicted from failure criteria using constant strain rate, creep, and stress relaxation measurements. For ABS-A the extrapolated creep and the constant strain rate failure criterion appeared to best predict the Tresca failure criterion pipe burst data. However, the hoop burst stress for ABS-N, adjusted with the von Mises failure criterion modification, was found to give the best agreement with the yield stress equivalent failure criterion for constant strain rate, creep, and stress relaxation. For ABS-N, the extrapolated creep failure criteria appeared to best predict the von Mises pipe burst failure criterion. In this study, the relationships between the failure criteria for the three different experimental techniques of constant strain rate, creep, and stress relaxation were shown to be reasonably interchangeable relative to a three-dimensional failure configuration such as pipe burst. Because this interchangeability approach was found to work so well in the laboratory, there is no reason to believe that this same technique, using the universal viscoelastic model, would not also work as well to predict failure criteria for design applications using finite-element analysis. Oil and gas transmission pipelines have a good safety record. This is due to a combination of good design, materials and operating practices; however, like any engineering structure, pipelines do occasionally fail. The most common causes of damage and failures in onshore and offshore, oil and gas transmission pipelines in Western Europe and North America are external interference (mechanical damage) and corrosion. Assessment methods are needed to determine the severity of such defects when they are detected in pipelines. Defects occurring during the fabrication of a pipeline are usually assessed against recognised and proven quality control (workmanship) limits. However, a pipeline will invariably contain larger defects at some stage during its life, and these will require a 'fitness-for-purpose' assessment to determine whether or not to repair the pipeline. Consequently, the past 40 years has seen the development of a number of methods for assessing the significance of defects. Some of these methods have been incorporated into industry guidance, others are to be found in the published literature. However, there is no definitive guidance that draws together all of the assessment techniques, or assesses each method against the published test data, or recommends best practice in their application.

This book describes a series of burst tests on model aluminium pipes containing dents and smooth gouges, specifically excluding cracks. The results show that gouges located near the axial extremity of a dent cause low-pressure failure, and indicate the presence of two regions of high strain at the axial extremities of the initial dent. For gouges in other areas of the dent, the Battelle flow stress dependent model for ductile failure of plain gouges in undented pipes is shown to give reasonable predictions of burst pressures, which are independent of the dent depth. In all of the pipes tested, failure was ductile, and this is attributable to the absence of cracks. Pipes withstood pressures in excess of the corresponding normal operating pressures, except where gouges were located near the axial extremity of the dent.

This handbook predicted the burst strength for defects that take the form of a large area of metal loss, where a continuum mechanics approach is more appropriate than a fracture mechanics approach.

— Gundel Wilson

Chapter 1

Introduction

Pipe (Fluid Conveyance)

A pipe is a tubular section or hollow cylinder, usually but not necessarily of circular cross-section, used mainly to convey substances which can flow-liquids and gases (fluids), slurries, powders, masses of small solids. It can also be used for structural applications; hollow pipe is far stiffer per unit weight than solid members.

In common usage the words *pipe* and *tube* are usually interchangeable, but in industry and engineering, the terms are uniquely defined. Depending on the applicable standard to which it is manufactured, pipe is generally specified by a nominal diametre with a constant outside diametre (OD) and a schedule that defines the thickness. Tube is most often specified by the OD and wall thickness, but may be specified by any two of OD, inside diametre (ID), and wall thickness. Pipe is generally manufactured to one of several international and national industrial standards. While similar standards exist for specific industry application tubing, tube is often made to custom sizes and a broader range of diametres and tolerances. Many industrial and government standards exist for the production of pipe and tubing. The term "tube" is also commonly applied to non-cylindrical sections, i.e., square or rectangular tubing. In general, "pipe" is the more common term in most of the world, whereas "tube" is more widely used in the United States.

Both "pipe" and "tube" imply a level of rigidity and permanence, whereas a *hose* (or hosepipe) is usually portable and flexible. Pipe assemblies are almost always constructed with the use of fittings such as elbows, tees, and so on, while tube may be formed or bent into custom configurations. For materials that are inflexible, cannot be formed or where construction is governed by codes or standards, tube assemblies are also constructed with the use of tube fittings.

Uses

- Domestic water systems
- Pipelines transporting gas or liquid over long distances
- Scaffolding
- Structural steel
- As components in mechanical systems such as:
    - o Rollers in conveyor belts
    - o Compactors (E.g.: steam rollers)
    - o Bearing casing
- Casing for concrete pilings used in construction projects
- High temperature or pressure manufacturing processes
- The petroleum industry:
    - o Oil well casing
    - o Oil refinery equipment
- Delivery of fluids, either gaseous or liquid, in a process plant from one point to another point in the process
- Delivery of bulk solids, in a food or process plant from one point to another point in the process
- The construction of high pressure storage vessels (note that large pressure vessels are constructed from plate, not pipe owing to their wall thickness and size).

Tube Drawing

Tube drawing is a metalworking process to size tube by shrinking a large diametre tube into a smaller one, by drawing the tube through a die. This process produces high-quality tubing with precise dimensions, good surface finish, and the added strength of cold working. Because it is so versatile, tube drawing is suitable for both large- and small-scale production. There are five types of tube drawing: tube sinking, mandrel drawing, stationary mandrel, moving mandrel, and floating mandrel. A mandrel is used in many of the types to prevent buckling or wrinkling in the workpiece.

Processes

Tube Sinking

Tube sinking, also known as free tube drawing, reduces the diametre of the tube without a mandrel inside the tube. The inner diametre (ID) is determined by the inner and outer diametre of the stock tube, the outer diametre of the final product, the length of the die landing, the amount of back tension, and the friction between the tube and the die. This type of drawing operation is the most economical, especially on thick-walled tubes and tubes smaller than 12 mm (0.47 in) in diametre, but does not give the best surface finish. As the tube thickness increases the surface finish quality decreases. This process is often used for the tubing on low cost lawn furniture.

Rod Drawing

Rod drawing is the process that draws the tube with a mandrel inside the tube; the mandrel is drawn with the tube. The advantage to this process is that the mandrel defines the ID and the surface finish and has a quick setup time for short runs. The disadvantages are that lengths are limited by the length of the mandrel, usually no more than 100 feet (30 m), and that a second operation is required to remove the mandrel, called *reeling*. This type of process is usually used on heavy walled or small ID tubes. Common applications include super-high pressure tubing and hydraulic tubing (with the addition of a finishing tube sinking operation). This process is also use for precision manufacturing of trombone handslides.

Fixed Plug Drawing

Fixed plug drawing, also known as stationary mandrel drawing, uses a mandrel at the end of the die to shape the ID of the tube. This process is slow and the area reductions are limited, but it gives the best inner surface finish of any of the processes. This is the oldest tube drawing method.

Floating Plug Drawing

Floating plug drawing, also known as floating mandrel drawing, uses a mandrel that is not anchored whatsoever to shape the ID of the tube. The mandrel is held in by the friction forces between the mandrel and the tube. This axial force is given by friction and pressure. The greatest advantage of this is that it can be used on extremely long lengths, sometimes up to 1,000 feet (300 m). The disadvantage is it requires a precise design otherwise it will give inadequate results. This process is often used for oil-well tubing.

Tethered Plug Drawing

Tethered plug drawing, also known as semi-floating mandrel drawing, is a mix between floating plug drawing and fixed plug drawing. The mandrel is allowed to float, but it still anchored via a tether. This process gives similar results to the floating plug process, except that it is designed for straight tubes. It gives a better inner surface finish than rod drawing.

Materials

Figure: Historic water mains from Philadelphia included wooden pipes

Pipe are made in many materials including ceramic, fibreglass, many metals, concrete and plastic. In the past wood and lead (Latin *plumbum*, from which comes the word 'plumbing') were commonly used.

Metallic pipes are commonly made from steel or iron; the finish and metal chemistry are peculiar to the use, fit and form. Typically metallic piping is made of steel or iron, such as unfinished, black (lacquer) steel, carbon steel, stainless steel or galvanised steel, brass, and ductile iron. Aluminium pipe or tubing may be utilised where iron is incompatible with the service fluid or where weight is a concern; aluminium is also used for heat transfer tubing such as in refrigerant systems. Copper tubing is popular for domestic water (potable) plumbing systems; copper may be used where heat transfer is desirable (i.e. radiators or heat exchangers). Inconel, chrome moly, and titanium steel alloys are used in high temperature and pressure piping in process and power facilities. When specifying alloys for new processes, the known issues of creep and sensitisation effect must be taken into account.

Lead piping is still found in old domestic and other water distribution systems, but it is no longer permitted for new potable water piping installations due to its toxicity. Many building codes now require that lead piping in residential or institutional installations be replaced with non-toxic piping or that the tubes' interiors be treated with phosphoric acid. According to a senior researcher and lead expert with the Canadian Environmental Law Association, "...there is no safe level of lead [for human exposure]".

Plastic tubing is widely used for its light weight, chemical resistance, non-corrosive properties, and ease of making connections. Plastic materials include polyvinyl chloride (PVC), chlorinated polyvinyl chloride (CPVC), fibre reinforced plastic (FRP), reinforced polymer mortar (RPMP), polypropylene (PP), polyethylene (PE), cross-linked high-density polyethylene (PEX), polybutylene (PB), and acrylonitrile butadiene styrene (ABS), for example. In many countries, PVC pipes account for most pipe materials used in buried municipal applications for drinking water distribution and wastewater mains.

Pipe may be made from concrete or ceramic, usually for low-pressure applications such as gravity flow or drainage. Pipes for sewage are still predominantly made from concrete or vitrified clay. Reinforced concrete can be used for large-diametre concrete pipes. This pipe material can be used in many types of construction, and is often used in the gravity-flow transport of storm water. Usually such pipe will have a receiving bell or a stepped fitting, with various sealing methods applied at installation.

Traceability and Positive Material Identification (PMI)

Maintaining the traceability between the material and this paperwork is an important quality assurance issue. QA often requires the heat number to be written on the pipe. Precautions must also be taken to prevent the introduction of counterfeit materials. As a backup to etching/labelling of the material identification on the pipe, positive material identification (PMI) is performed using a handheld device; the device scans the pipe material using an emitted electromagnetic wave (x-ray fluorescence/XRF) and receives a reply that is spectrographically analysed.

Nominal Pipe Size

Nominal Pipe Size (NPS) is a North American set of standard sizes for pipes used for high or low pressures and temperatures. Pipe size is specified with two non-dimensional numbers: a nominal pipe size (NPS) for diametre based on inches, and a schedule (Sched. or Sch.) for wall thickness. NPS is often incorrectly called National Pipe Size, due to confusion with national pipe thread (NPT). The European designation equivalent to NPS is *DN* (diamètre nominal/nominal diametre/Durchmesser nach Norm), in which sizes are measured in millimetres. The term NB (nominal bore) is also frequently used interchangeably with NPS.

History

In March 1927, the American Standards Association authorised a committee to standardise the dimensions of wrought steel and wrought iron pipe and tubing. At that time only a small selection of wall thicknesses were in use: standard weight (STD), extra-strong (XS), and double extra-strong (XXS), based on the iron pipe size (IPS) system of the day. However these three sizes did not fit all applications. The committee surveyed the industry and created a system of schedule numbers that designated wall thicknesses based on smaller steps between sizes, although IPS and NPS numbers remain equivalent.

The original intent was that each schedule would relate to a given pressure rating, however the numbers deviated so far from wall thicknesses in common use that this original intent could not be accomplished. Also, in 1939, it was hoped that the designations of STD, XS, and XXS would be phased out by schedule numbers, however those original terms are still in common use today (although sometimes referred to as standard, extra-heavy (XH), and double extra-heavy (XXH), respectively). Since the original schedules were created, there have been many revisions and additions to the tables of pipe sizes based on industry use and on standards from API, ASTM, and others.

Stainless steel pipes, which were coming into more common use in the mid 20th century, permitted the use of thinner pipe walls with much less risk of failure due to corrosion. By 1949 thinner schedules 5S and 10S, which were based on the pressure requirements modified to the nearest BWG number, had been created, and other "S" sizes followed later. Due to their thin walls, the smaller "S" sizes can not be threaded together according to ASME code, but must be fusion welded.

Application

Based on the NPS and schedule of a pipe, the pipe outside diametre (OD) and wall thickness can be obtained, which are based on ASME standards B36.10M and 5.19M. For example, NPS 14 Sch 40 has an

OD of 14 inches and a wall thickness of 0.437 inches. However the NPS and OD values are not always equal, which can create confusion.

- For NPS 1/2 to 12 inches, the NPS and OD values are different. For example, the OD of an NPS 12 pipe is actually 12.75 inches. To find the actual OD for each NPS value.
- For NPS 14 inches and up, the NPS and OD values are equal. In other words, an NPS 14 pipe is actually 14 inches OD.

The reason for the discrepancy for NPS [! to 12 inches is that these NPS values were originally set to give the same *inside* diametre (ID) based on wall thicknesses standard at the time. However, as the set of available wall thicknesses evolved, the ID changed and NPS became only indirectly related to ID and OD.

For a given NPS, the OD stays fixed and the wall thickness increases with schedule. For a given schedule, the OD increases with NPS while the wall thickness stays constant or increases. Using equations and rules in ASME B31.3 Process Piping, it can be shown that pressure rating decreases with increasing NPS and constant schedule. Some specifications use pipe schedules called standard wall (STD), extra strong (XS), and double extra strong (XXS), although these actually belong to an older system called iron pipe size (IPS). The IPS number is the same as the NPS number. STD is identical to SCH 40S, and 40S is identical to 40 for NPS 1/8 to NPS 10, inclusive. XS is identical to SCH 80S, and 80S is identical to 80 for NPS 1/8 to NPS 8, inclusive. XXS wall is thicker than schedule 160 from NPS 1/8" to NPS 6" inclusive, and schedule 160 is thicker than XXS wall for NPS 8" and larger. The “S” designation, for example “NPS Sch 10S”, most often indicates stainless steel pipes. However some stainless steel pipes are available in steel designations, so strictly speaking the “S” designation only differentiates B36.19M pipe from B36.10M pipe. Both polyvinyl chloride pipe (PVC) and chlorinated polyvinyl chloride pipe (CPVC) are made in NPS sizes.

Standards

The manufacture and installation of pressure piping is tightly regulated by the ASME “B31” code series such as B31.1 or B31.3 which have their basis in the ASME Boiler and Pressure Vessel Code. This code has the force of law in Canada and the USA. Europe and the rest of the world has an equivalent system of codes. Pressure piping is generally pipe that must carry pressures greater than 10 to 25 atmospheres, although definitions vary. To ensure safe operation of the system, the manufacture, storage, welding, testing, etc. of pressure piping must meet stringent quality standards.

Manufacturing standards for pipes commonly require a test of chemical composition and a series of mechanical strength tests for each heat of pipe. A heat of pipe is all forged from the same cast ingot, and therefore had the same chemical composition. Mechanical tests may be associated to a lot of pipe, which would be all from the same heat and have been through the same heat treatment processes. The manufacturer performs these tests and reports the composition in a mill traceability report and the mechanical tests in a material test report, both of which are referred to by the acronym MTR. Material with these associated test reports is called traceable. For critical applications, third party verification of these tests may be required; in this case an independent lab will produce a certified material test report (CMTR), and the material will be called certified.

Some widely used pipe standards are:

- The API range-now ISO 3183. E.g.: API 5L Grade B-now ISO L245 where the number indicates yield strength in MPa
- ASME SA106 Grade B (Seamless carbon steel pipe for high temperature service)
- ASTM A312 (Seamless and welded austenitic stainless steel pipe)
- ASTM C76 (Concrete Pipe)
- ASTM D3033/3034 (PVC Pipe)
- ASTM D2239 (Polyethylene Pipe).

API 5L was changed in the second half of 2008 to edition 44 from edition 43 to make it identical to ISO 3183. It is important to note that the change has created the requirement that sour service, ERW pipe, pass a hydrogen induced cracking (HIC) test per NACE TM0284 in order to be used for sour service.

- ACPA [American Concrete Pipe Association]
- AWWA [American Water Works Association]
- AWWA M45.

Pipefitter

A pipefitter (also called steamfitter) is a tradesman who lays out, assembles, fabricates, maintains and repairs mechanical piping systems. Pipefitters usually go through a mix of apprentice and trade school training. Journeyman pipefitters/steamfitters deal with industrial process piping and heating/cooling systems. Typical industrial process pipe is under high pressure, which requires metals such as carbon steel, stainless steel, and many different alloy metals fused together through precisely cutting, threading, grooving, bending and welding, whereas a plumber concentrates on piping systems for heating/cooling as well as utility liquids potable water, sewage, drains, etc. in the industrial/commercial/institutional or residential atmosphere operating at lower pressures. Utility piping typically consists of copper, PVC, CPVC, polyethylene, and galvanised pipe which is typically glued, soldered, or threaded. Other types of piping systems include steam, ventilation, hydraulics, chemicals or fuel.

Figure: Large-scale piping system in an HVAC mechanical room

In the United States, many states require pipefitters to be licensed. Requirements differ from state to state but most include a four to five year apprenticeship and testing before becoming a licensed journeyman.

Occupational Summary

Pipefitter Skills include:

- Layout
- Blueprint reading
- Detailing

- CAD Drawing Coordinators
- Pipe Threading
- Pipe Grinding
- Welding (MiG, GMAW, TiG, SMAW, Orbital)
- Plasma cutting
- Mechanical Pipe cutting
- Gas Arc cutting
- Rigging
- Brazing
- Soldering
- Mitring
- Tube Bending
- Valve installation and repair
- Preparation and installation of medical gas piping.

Employers that hire pipefitters include:

- Heating, ventilation and air conditioning HVAC construction firms
- Fire fighting pipelines
- Mechanical construction firms
- Pulp mills and paper mills
- Primary steel producers
- Utility companies
- Motor vehicle manufacturers
- Residential building developers
- Fabrication metal Metal fabricating companies
- Oil refineries
- Chemical plants
- Pharmaceutical companies
- Shipyards and ship repair facilities
- Nuclear Power Plants
- Coal-Burning Power Plants
- Gas Plants
- Breweries
- Semiconductor plants

- Service companies that maintain and install controls such as thermostats and balancing valves
- Hospitals and healthcare facilities
- Glass manufacturing Plants.

Recommended or required skills include physical strength and manual dexterity, reading and interpreting blueprints and specifications; making detailed sketches for pipe and equipment fabrication and installation; cutting openings for pipe using various hand or power tools; measuring, cutting with blades and torches, threading, grooving, bending, soldering, brazing, welding, assembling, mitering, joining and installing pipes, valves and fittings.

Trade Groups

In North America, Union pipefitters are members of the United Association. Wages vary from area to area, based on contracts between the local union and contractors.

Pipe Supports

Pipes are usually either supported from below or hung from above (but may also be supported from the side), using devices called pipe supports. Supports may be as simple as a pipe "shoe" which is akin to a half of an I-beam welded to the bottom of the pipe; they may be "hung" using a clevis, or with trapeze type of devices called pipe hangers. Pipe supports of any kind may incorporate springs, snubbers, dampers, or combinations of these devices to compensate for thermal expansion, or to provide vibration isolation, shock control, or reduced vibration excitation of the pipe due to earthquake motion. Some dampers are simply fluid dashpots, but other dampers may be active hydraulic devices that have sophisticated systems that act to dampen peak displacements due to externally imposed vibrations or mechanical shocks. The undesired motions may be process derived (such as in a fluidized bed reactor) or from a natural phenomenon such as an earthquake (design basis event or DBE).

Pipe hanger assembles are usually attached with pipe clamps. Possible exposure to high temperatures and heavy loads should be included when specifying which clamps are needed.

Piping and Plumbing Fittings

Fittings are used in pipe and plumbing systems to connect straight pipe or tubing sections, to adapt to different sizes or shapes, and for other purposes, such as regulating or measuring fluid flow. The term *plumbing* is generally used to describe conveyance of water, gas, or liquid waste in ordinary domestic or commercial environments, whereas *piping* is often used to describe high-performance (e.g. high pressure, high flow, high temperature, hazardous materials) conveyance of fluids in specialised applications. The term *tubing* is sometimes used for lighter-weight piping, especially types that are flexible enough to be supplied in coiled form. Fittings (especially uncommon types) require money, time, materials, and tools to install, so they are a non-trivial part of piping and plumbing systems. Valves are technically fittings, but are usually discussed separately.

Standards

There are certain standard codes that need to be followed while designing or manufacturing any piping system. Organisations that promulgate piping standards include:

- ASME - The American Society of Mechanical Engineers
- ASTM - American Society for Testing and Materials
- API - American Petroleum Institute
- AWS - American Welding Society
- AWWA - American Water Works Association
- MSS – Manufacturers' Standardisation Society
- ANSI - American National Standards Institute
- NFPA - National Fire Protection Association
- EJMA - Expansion Joint Manufacturers Association.

For example pipes need to conform to the dimensional requirements of :

- ASME B36.10M - Welded and Seamless Wrought Steel Pipe
- ASME B36.19M - Stainless Steel Pipe
- ASME B31.3 2008 - Process Piping.

The B31.3 code has requirements for piping found in petroleum refineries; chemical, pharmaceutical, textile, paper, semiconductor, and cryogenic plants; and related processing plants and terminals. This code specifies requirements for materials and components, design, fabrication, assembly, erection, examination, inspection, and testing of piping. This Code is applicable to piping for all fluids including: (1) raw, intermediate, and finished chemicals; (2) petroleum products; (3) gas, steam, air and water; (4) fluidized solids; (5) refrigerants; and (6) cryogenic fluids.

Materials

The material with which a pipe is manufactured often forms as the basis for choosing any pipe. Materials that are used for manufacturing pipes include:

- Carbon Steel (CS)
- Low Temperature Service Carbon Steel (LTCS)
- Stainless Steel (SS)
- Non-Ferrous Metals (Inconel, Incoloy, Cupro-nickel etc.)
- Non-Metallic (ABS, GRE, PVC, HDPE, tempered glass, etc.)
- Chrome-molybdenum steel (Alloy steel) — Generally used for high temperature service.

The bodies of fittings for pipe and tubing are most often of the same base material as the pipe or tubing being connected, for example, copper, steel, polyvinyl chloride (PVC), chlorinated polyvinyl chloride (CPVC), or acrylonitrile butadiene styrene (ABS). However, any material that is allowed by the plumbing, health, or building code (as applicable) may be used, but must be compatible with the other materials in the system, the fluids being transported, and the temperatures and pressures inside and outside of the system. For example, brass- or bronze-bodied fittings are common in otherwise copper piping and plumbing systems. Fire hazards, earthquake resistance, and other factors also influence choice of fitting materials.

Gaskets

Gaskets are used for sealing of flange joints. In general, gaskets should not be reused. Various types of gaskets are available depending upon their construction, materials, and features. The following are the type of gaskets commonly used:

- Non-Metallic Gaskets (ASME B 16.21)
- Spiral-Wound Gaskets (ASME B 16.20)
- Ring Joint Gaskets (ASME B 16.20).

Non-Metallic Gaskets are used with flat face or raised face flanges. Spiral-Wound Gaskets are used with raised face flanges. They are available with an inner ring and outer ring, which is also known as the cantering ring. Ring Joint Gaskets are used with Ring Type Joint (RTJ) flanges. They are available in octagonal or oval cross sections.

A very high surface stress is developed between an RTJ gasket and the flange groove when RTJ is bolted up in a flange. This leads to plastic deformation of the gasket. Thus, the hardness of the gasket is kept less than the hardness of the groove to achieve coining i.e bringing two metal surfaces of different hardness so tightly together that the softer surface deforms to match harder surface exactly in shape and finish. Hence RTJ gaskets are recommended for reuse. Gender of connectors and fasteners

Figure: A male threaded pipe, and female threaded elbow.

In electrical and mechanical trades and manufacturing, each half of a pair of mating connectors or fasteners is conventionally assigned the designation male or female. The "female" connector is generally a receptacle that receives and holds the "male" connector.

The assignment is by direct analogy with genitalia and sexual intercourse; the part bearing one or more protrusions, or which fits inside the other, being designated male in contrast to the part containing the corresponding indentations, or fitting outside the other, being designated female. Extension of the analogy results in the verb to mate being used to describe the process of connecting two corresponding parts together.

In some cases (notably electrical power connectors), the gender of connectors is selected according to rigid rules, to enforce a sense of one-way directionality (e.g. a flow of power *from* one device *to* another). This gender distinction is implemented to enhance safety or ensure proper functionality by preventing unsafe or non-functional configurations from being set up.

In terms of mathematical graph theory, an electrical power distribution network made up of plugs and sockets is a directed tree, with the directionality arrows corresponding to the female-to-male transfer of electrical power through each mated connection. This is an example where male and female connectors have been deliberately designed and assigned to physically enforce a safe network topology.

In other contexts, such as plumbing, one-way flow is *not* enforced through connector gender assignment. Flows through piping networks can be bidirectional, as in underground water distribution networks which have designed-in redundancy. In plumbing situations where one-way flow *is* desired, it is implemented through other means (e.g. gravity flow, one-way check valves), and not through male-female gender schemes.

Mechanical Fasteners

Figure: Female nut threaded onto a male bolt

In mechanical design, the prototypical "male" component is a threaded bolt, but an alignment post, a mounting boss, or a sheet metal tab connector can also be considered as male. Correspondingly, a threaded nut, an alignment hole, a mounting recess, or sheet metal slot connector is considered to be female.

While some mechanical designs are "one-off" custom setups not intended to be repeated, there is an entire fastener industry devoted to manufacturing mass-produced or semi-custom components. To avoid unnecessary confusion, conventional definitions of fastener gender have been defined and agreed upon.

Modular Construction Toys

Figure: Lego toy brick connections are male on top, and female underneath

Although this aspect is not highlighted in their promotional literature, several common construction toys embody gendered (and in some cases, genderless) mechanical interconnections. This should not be surprising, since these toys feature the nearly infinite flexibility and versatility of shape that a modular interconnect architecture can enable. Mathematicians have begun to classify well-known construction sets using group theory to study the combinatoric possibilities of structures that can be built.

For example, the canonical LEGO plastic blocks have "female" indentations on the lower surface, and "male" bosses or protrusions on the upper surfaces. Meccano and Erector sets are rife with gendered connections, starting with the nut-and-bolt fasteners they use in profusion.

Stickle bricks, using interdigitated plastic protrusions, are effectively genderless. Lincoln logs use a very simple form of genderless connections. Kapla or KEVA planks are extremely simple genderless systems interconnected only by gravity.

Plumbing Connectors

In plumbing fittings, the "M" or "F" usually comes at the beginning rather than the end of the abbreviated designation. For example:

- MIP denotes Male International Pipe thread;
- FIP denotes Female International Pipe thread.

A short length of pipe having an MIP thread at both ends is sometimes called a nipple. A short pipe fitting having an FIP thread at both ends is sometimes called a coupling.

Hermaphroditic connections, which include both male and female elements in a single unit, are used for some specialised tubing fittings, such as Storz firehose connectors.

Downspout Connectors

Downspouts (downpipes, rain conductors or leaders) are used to convey rainwater from roof gutters to the ground through hollow pipes or tubes. These tubes usually come in sections, joined by inserting the male end (often crimped with a special tool to slightly reduce its size) into the female end of the next section. These connections are usually not sealed or caulked, instead relying on gravity to move the rainwater from the male end and into the receiving female connection located directly below.

Ductwork Connectors

Sheet metal ductwork for conveying air in HVAC systems typically uses gendered connections. Typically, the airflow through a ductwork connection is from male to female. However, since one-way flow is implemented by forced-air fans or blowers, "backwards" gendered connections can be seen frequently in some systems, since all connections are typically sealed with duct sealing mastic or tape to prevent leakage anyway. The flow convention is usually loosely adhered to for simplicity of design, and to reduce the number of gender changer fittings required, but exceptions are made whenever expedient.

Electrical and Electronic Connectors

Figure: A short adaptor cable, with male *plug* on left, and female *plug* on right

Although the gender of tubing and plumbing fittings is usually obvious, this may not be true of electrical connectors because of their more complex and varying constructions. Instead, connector gender is conventionalised and thus can be somewhat obscure to the uninitiated.

For example, the female D-subminiature connector body projects outward from the mounting plane of the chassis, and this protrusion could be erroneously construed as male. Instead, the "maleness" of the D-subminiature connectors is defined by specific presence of *male pins*, rather than by the protrusion of the connector, which is also true for many other pin-based connectors like XLR. The male/female distinction is more obvious with ring crimp lug connectors which are placed around a screw post, but again with spade or split ring crimp lug connectors the end alone is not obviously female.

Further confusion can be caused by the term "jack", which is used for both female and male connectors and typically refers to the fixed (panel) side of a connector pair. IEEE STD 100 and ANSI Y32.16 (identical to IEEE 200-1975 and replaced by ASME Y14.44-2008) define "Plug" and "Jack" by location or mobility, rather than gender. A connector in a fixed location is a jack, and a moveable connector is a plug.

The distinction is relative, so a portable radio is considered stationary compared to the cable from the headphones; the radio has a jack, and the headphone cable has a plug. Where the relationship is equal, such as when two flexible cables are connected, each is considered a plug and annotated with the P reference designator.

It is common practice to use female connectors for jacks, so the informal gender-based usage often happens to agree with the functional description of the technical standards. However, this is *not* always the case; often-seen exceptions include a computer's AC Power Inlet and EIA232 DE9 Serial Port, or the male coaxial power jacks for connecting external power adapters to portable equipment.

To summarise, it is considered best practice to use "male" and "female" for connector *gender*, and "plug" and "jack" for connector *function* or *mobility*.

Non-standard Usage

In much of the non-English-speaking world, the word "jack" may refer to the *plug* on the end of a removable cable; this can be quite surprising. Apparently, these connectors were originally referred to as "jack plugs", or plugs intended to be mated with fixed receptacles (what English speakers would call "jacks"), but the second word was dropped. This variant usage is in direct contradiction to common usage and official standards in predominantly English-speaking countries, such as the US, Canada, UK, and Australia.

Abbreviations and Alternate Terminology

The standard letters "M" and "F" are commonly used in part numbers to designate connector gender. For example, in Switchcraft XLR microphone or hydrophone connectors, the part numbers are denoted as follows:

- A3F = Audio 3-pin Female connector;
- A3M = Audio 3-pin Male connector.

The terms plug, pin, and prong are also often used for "male" connectors, and receptacle, socket, and slot are used for "female" connectors. In many cases these terms are more common than *male* and *female*, especially in documentation intended for the non-specialist. These nearly synonymous terms can cause a fair amount of confusion when the designations are shortened in labels. For example, a female high-density D-subminiature connector with a size 1 shell can be named DE15F or DE15S. Both terms mean the same thing but could be construed to be completely different items. Similarly, a male standard-density D-sub with a size 1 shell can be named DE9M or DE9P; a female standard-density D-sub with a size 2 shell can be named DA15F or DA15S; a male high-density D-sub with a size 3 shell can be named DB44M or DB44P; and so forth.

Gender Selection in Electronic Design

Electronic designers often select female jack connectors for fixed mounting on electronic equipment they design. This is usually done because female connectors are more resistant to damage or contamination, by virtue of their concealed or recessed electrical contacts. A damaged motherboard connector can result in the scrapping of an expensive piece of electronic equipment. The risk of expensive damage is reduced by relegating the more exposed male contacts to connecting cables, which can be repaired or replaced at lower cost.

Such cost and reliability considerations probably drove the design decision to use female jack connectors on many computer terminals (and some personal computers) for the serial port, in direct violation of the connector gender convention specified in the RS-232 standard for "DTE" computer equipment.

This confusing reversal of the RS-232 connector gender convention would cause many hours of frustration for ill-informed end users, as they tried to troubleshoot non-functional serial port equipment connections.

In the case of electrical power connections, designers dare not to reverse connector gender in such a casual fashion, since such misuse of AC line power connectors is very unsafe and even illegal.

Fortunately, a special male IEC 60320 C14 connector is available which is recessed below the surface of a mounting panel, providing the desired protection from mechanical damage while conforming to safety regulations.

Safety and Electrical Connector Gender

In electrical connections where voltage or current is sufficient to cause injury, the part connected to the power *source* is invariably female, with concealed contacts, to prevent inadvertent touching of live conductors. A male plug, with fully exposed protruding contacts, is installed on the cord of (or directly onto) the device *receiving* the power.

In the case of consumer-level AC power, connector gender is used to implicitly enforce safe use of power connectors. Because of this consideration, it is illegal under electrical code to make or use any *gender changers* to connect AC line power to consumer-level equipment.

Figure: Electrical power outlets are female for safety.

Figure: Common 5.5×2.5 mm coaxial power connectors. Power is provided by the female *plug* on the right to the male *jack* on the left; the exposed conductors are not hazardous due to the low voltage.

In low-voltage use such as for data communications, electrical shock hazard is not an issue, and male or female connectors are used based on other engineering factors such as convenience of use, cost, or ease of manufacturing. For example, the common "patch cables" used for Ethernet (and the similar cords used for telephones) typically have male modular plugs on both ends, to connect to jacks on equipment or mounted in walls.

As an illustrative example of some design tradeoffs in power connector selection, consider the picture to the left. A commonly seen coaxial power connector is usually set up so that power is fed *from* the female plug on the right *into* the male jack on the left (which is typically a part of the electronic device accepting the power). Although the plug is female, with a partially recessed centre contact, it is still possible for casual accidental contact with a metallic object to short-circuit the power source. Depending on the design of the power adapter, it may react to a short circuit by shutting down temporarily, or instead by blowing out an internal safety fuse.

In this example, the marginal reliability of the connector choice was deemed to be acceptable by the equipment designer, since the power adapter supplies low voltage that does not pose an electric shock hazard. The potential fire hazard from accidental short-circuiting is addressed by the internal safety fuse, although this requires that a failed power adapter must be completely replaced. In a different design, if the power adapter were intended to supply a voltage sufficient to cause electrical shock, the semi-exposed centre contact of the female plug would be considered unacceptably hazardous, requiring a different choice of power connector.

Ambiguous Gender Assignment

As a side issue, some electrical connectors are hermaphroditic because they include both male and female elements in a single unit intended to interconnect freely, without regard for gender.

As an additional complication, certain electronic connector designs may incorporate combinations of male and female pins in a *single* connector body, for mating with a complementary connector with opposite gender pins in corresponding positions.. In these unusual cases, gender is often defined by the shape of the connector *body*, rather than the mixed-gender connector *pins* and *sockets*. These types of connectors are not strictly speaking hermaphroditic, since mating connectors are *not* freely interchangeable. An informal term that has been used for these connectors is “bisexual”, in addition to the more official terminology, mixed-gender. Thus, for example, one can have a mixed-gender female plug that connects to a mixed-gender male jack (though a reversed gender assignment of connectors would be a more typical design choice in this example).

Male connector pins are often protected by a shell (also called a shroud, surround, or shield), which may envelop the entire female connector when mated. RF connectors often have multiple layers of interlocking shells to properly connect the shields of coaxial and triaxial cable. In such cases, the gender is assigned based on the innermost connecting point. Another ambiguous situation arises with the connectors used for USB, FireWire (IEEE-1394), HDMI, and Thunderbolt serial data bus connections. Close examination of these connectors reveals that the contact “pins” are not actually pins, but instead are conductive surfaces that slide past each other when they mate. Therefore, the traditional pin and socket nomenclature is not applicable. Instead, most computer hardware people fall back to referring to the wrap-around metal shield on the plug connector *as if* it were a connector *pin*. By this convention, the connectors on serial bus cables are "male plugs", and the corresponding connectors on equipment are "female jacks".

A casual glance at a USB "Type A" plug connector may give the false impression that it is hermaphroditic. However, a physical attempt to mate two USB "Type A" cables with each other reveals the fact that the connectors will not interconnect (at least not without interposing the appropriate USB gender changer). Classifying according to mathematical graph theory, USB buses are directed trees, whereas FireWire buses have a true bus network topology. This difference is reflected in the bus connectors used, in that USB cables are asymmetrical (one end Type A, other end Type B) while FireWire cables may have identical connectors at both ends.

Genderless (Hermaphroditic) Connectors

Figure: Paired "knuckle" type hermaphroditic mechanical couplers for railcars (viewed from above

Figure: Arduino "shield" boards connected via stackable bus connectors

By definition, a hermaphroditic connector includes mating surfaces having simultaneous male and female aspects, involving complementary paired identical parts each containing both protrusions and indentations. These mating surfaces are mounted into identical fittings which can freely mate with any other, without regard for gender (provided that the size and type are already matched). Alternative names include hermaphrodite, genderless, sexless, combination (or combo), two-in-one, two-way, and other descriptive terms. Several of these latter alternate names are ambiguous in meaning, and should not be used unless carefully defined in context. True hermaphroditic connectors should not be confused with *mixed gender* connectors, which are described elsewhere in this article.

Another closely related type is the stackable connector for electronics, which typically has male pins on one surface, and complementary female sockets on the opposite surface, allowing multiple units to be stacked up like plastic milk crates. Examples of this include stackable banana plugs, and interconnect cables specified for the IEEE-488 instrumentation bus. Stackable mezzanine bus connectors are used on some modular microcomputer accessory boards for systems such as the Arduino add-on daughterboards called "shields". The older PC/104 embedded PC modules use a similar stackable format for interconnection. Stackable connectors are not classified as hermaphroditic in the strictest sense, but are often described as such in looser usage.

The hermaphroditic design is useful when multiple complex or lengthy components must be arbitrarily connected in various combinations.

For example, if hoses have hermaphroditic fittings, they can be connected without having to pull a lengthy hose and reverse it because it has the wrong gender to connect to another hose. Some military fibre optical cables also have hermaphroditic connectors to prevent "wrong gender" connector problems in field deployments. In a similar fashion, railcars are usually equipped with hermaphroditic railway coupling mechanisms that allow either end of the vehicle to be connected to a train consist without having to turn the railcar around first.In the absence of genderless connectors, *gender changer* fittings might be used to enable certain connections. The designer of a connection system may use one or both schemes to allow arbitrary connectivity, or even combine both schemes into a single system.

When an enforced sense of unidirectionality or "one-way flow" is required for safety or other reasons (for example, AC electrical power connections), a strict assignment of connector genders is implemented to prevent undesired configurations, and gender changers are banned.

Some commonly seen examples of hermaphroditic connectors include the SAE connector for 12VDC power, jackhammer air hose connectors, and the Anderson Powerpole series of modular high-current power connectors. The IBM token ring connector was another widespread example, but it has become obsolete and is being phased out. The General Radio Corporation (GenRad) developed a hermaphroditic coaxial radio frequency connector often called the "GR connector".

Some audio multicore cables are fitted with hermaphroditic multipin quick-disconnect connectors for ease of use in the field. One style of this audio signal cable is fitted on both ends with connectors that are each populated half with pins and half with sockets. The advantage to the user is that it does not matter which end connects to the stage and which to the audio mixer, facilitating faster set up. Another style of connector uses hybrid male/female pins with a receiving slot fitted in the centre of each two-tine pin, and relies on 90-degree rotation of the pin axes to mate. The connector housings themselves are sexed male and female.

Gender Changers

Devices used for mating two connectors of the same gender have a wide variety of terms, including for example: "gender changer", "gender mender", "gender bender", "gender blender", "sex changer", and "homosexual adapter". A specific gender changer can be referred to by either the gender of its connectors, or the gender which it is designed to connect to, resulting in a thoroughly ambiguous terminology. Thus a "male gender changer" might have female connectors to mate two male ends, or male connectors to mate two female ends.

Examples

- A power cord on an appliance terminates in a (male) plug; it connects to a (female) socket in a wall or on an extension cord.
- Coaxial cables used for video or other high-frequency signals are normally terminated, at both ends, in a connector comprising an inner pin and an outer fixed or rotating shell; these are conventionally reckoned as male.
- A threaded nut is female and a bolt is male.
- Connectors for air brake hoses on heavy trucks and railroad

equipment use genderless "gladhand" connectors. In railroad air brake use, this makes the orientation of rolling stock irrelevant, and is used with the standard North American railroad coupler that connects cars together, also genderless.

- Piping and plumbing fittings
- The Talmud (c. 500 CE) describes arrow heads and mating shafts as being male and female, i.e. a prong on a male arrow head fits into a hollowed out shaft and vice versa.

Common Fittings for Both Piping and Plumbing

While there are hundreds of specialised fittings manufactured, some common types of fittings are used widely in piping and plumbing systems

Street Elbow

A street elbow (sometimes called a street ell or service ell) is a type of plumbing or piping fitting intended to join a piece of pipe and another fitting at an angle. The difference between a street elbow and a regular elbow is the nature of the connections on either end.

A regular elbow has a *female* hub or FIP connection on *both* ends. A hub is a female ABS, PVC or copper pipe receptacle; FIP denotes "Female Iron Pipe" threads, that is, it receives threaded MIP iron, brass or plastic pipe on both ends. FIP is also known as FPT, *Female Pipe Thread*, and MIP is also known as MPT, *Male Pipe Thread.* Collectively, these are known as National Pipe Thread (NPT) fittings.

Instead, a street elbow has a *female* fitting (hub or FIP) on one end and a *male* pipe or MIP fitting on the other end. The advantage of the street elbow is that it can be connected directly to another fitting without having to use an additional short connecting piece called a pipe nipple. Street elbows are available in 90°, 45° and 22.5° bends. They can be used in water supply, drainage, sewers, vents, central vacuum systems, compressed air and gas lines, HVAC, sump pump drains, and any location where plumbing fittings would be used to join sections of pipe.

Coupling

A coupling connects two pipes to each other. If the size of the pipe is not the same, the fitting may be called a reducing coupling or reducer, or an adapter. By convention, the term "expander" is not generally used for a coupler that increases pipe size; instead the term "reducer" is used.

Figure: Pipe coupling (copper sweat)

Union

A union is similar to a coupling, except it is designed to allow quick and convenient disconnection of pipes for maintenance or fixture replacement. While a coupling would require either solvent welding, soldering or being able to rotate with all the pipes adjacent as with a threaded coupling, a union provides a simple transition, allowing easy connection or disconnection at any future time. A standard union pipe is made in three parts consisting of a nut, a female end, and a male end. When the female and male ends are joined, the nuts then provide the necessary pressure to seal the joint. Since the mating ends of the union are interchangeable, changing of a valve or other device can be achieved with a minimum loss of time. Pipe unions are essentially a type of flange connector, as discussed further below.

Figure: A combination pipe union and reducer fitting (brass threaded)

In addition to a standard union, there exist dielectric unions which are used to separate dissimilar metals (such as copper and galvanised steel) to avoid the damaging effects of galvanic corrosion. When two dissimilar metals are in contact with an electrically conductive solution (even tap water is conductive), they will form a battery and generate a voltage by electrolysis. When the two metals are in direct contact with each other, the electric current from one metal to the other will cause a movement of ions from one to the other, dissolving one metal and depositing it on the other. A dielectric union breaks the electric current path with a plastic liner between two halves of the union, thus limiting galvanic corrosion.

Reducer

A reducer is the component in a pipeline that reduces the pipe size from a larger to a smaller bore (inner diametre).

The length of the reduction is usually equal to the average of the larger and smaller pipe diametres. There are two main types of reducer: concentric and eccentric reducers. A reducer can be used either as nozzle or as diffuser depending on the mach number of the flow.

Olets

Whenever branch connections are required in size where reducing tees are not available and/or when the branch connections are of smaller size as compared to header size, olets are generally used. The following are few configurations of olet connections :

- Flanged Olet
- Socket-Weld & Threaded Olet
- Lateral & Elbow Olets
- Nipple Olet
- Butt-Weld Olet
- Swage Nipples.

Tee

Figure: Pipe tee (copper sweat)

A tee is the most common pipe fitting. It is available with all female thread sockets, all solvent weld sockets, or with opposed solvent weld sockets and a side outlet with female threads. It is used to either combine or split a fluid flow. It is a type of pipe fitting which is T-shaped having two outlets, at 90° to the connection to the main line. It is a short piece of pipe with a lateral outlet. A tee is used for connecting pipes of different diametres or for changing the direction of pipe runs. They are made of various materials and available in various sizes and finishes. They are extensively used in pipeline networks to transport two-phase fluid mixtures. They are categorised as:

- Equal
- Unequal

When the size of the branch is same as header pipes, equal tee is used and when the branch size is less than that of header size, reduced tee will be used. Most common are tees with the same inlet and outlet sizes. Some of the industrial tees are Straight Tee, Reducing Tee, Double Branch Tee, Double Branch Reducing Tee, Conical Tee, Double Branch Conical Tee, Bullhead Tee, Conical Reducing Tee, Double Branch Conical Reducing Tee, Tangential Tee, and Double Branch Tangential Tee. The above tees are categorised on the basis of their shapes and structure. They can also be classified on the basis of the application they are required to perform.

The three outlet sizes should be named in order (e.g. left, middle, right; measuring 15-22-15).

Cross

Cross fittings are also called 4-way fittings. If a branch line passes completely through a tee, the fitting becomes a cross. A cross has one inlet and three outlets, or vice versa. They often have solvent welded socket ends or female threaded ends. Cross fittings can generate a huge amount of stress on pipe as temperature changes, because they are at the centre of four connection points. A tee is more steady than a cross, as a tee behaves like a three-legged stool, while a cross behaves like a four-legged stool. (Geometrically, "any 3 non-colinear points define a plane" thus 3 legs are inherently stable.) Crosses are common in fire sprinkler systems, but not in plumbing, due to their extra cost as compared to using two tees.

Cap

A type of pipe fitting, usually liquid or gas tight, which covers the end of a pipe. A cap is used like plug, except that the pipe cap screws or attaches on the male thread of a pipe or a nipple. A cap may have a solvent weld socket end or a female threaded end and the other end closed off. If a solvent weld cap is used to provide for a future connection point, several inches of pipe must be left before the cap.

This is because when the cap is cut off for the future connection, enough pipe must remain to allow a new fitting to be glued onto it. In plumbing systems that use threads, the cap has female threads. Industrial caps can be round, square, rectangular, U-shaped, I-shaped and have a round hand grip or a flat hand grip.

Figure: Pipe cap (copper sweat)

Plug

A plug closes off the end of a pipe. It is similar to a cap but it fits inside the fitting it is mated to. In a threaded iron pipe plumbing system, plugs have male threads. Some of the popular types of plugs are:

- Mechanical pipe plug
- Pneumatic disk pipe plug
- Single size pneumatic all rubber pipe plug
- Multi-size pneumatic pipe plug
- Multi-size flow-through pipe plug
- High pressure pipe plug.

Nipple

A short stub of pipe, usually threaded steel, brass, chlorinated polyvinyl chloride (CPVC) or copper; occasionally just bare copper. A nipple is defined as being a short stub of pipe which has external male pipe threads at each end, for connecting two other fittings. Nipples are commonly used for plumbing and hoses, and second as valves for funnels and pipes.

Barb

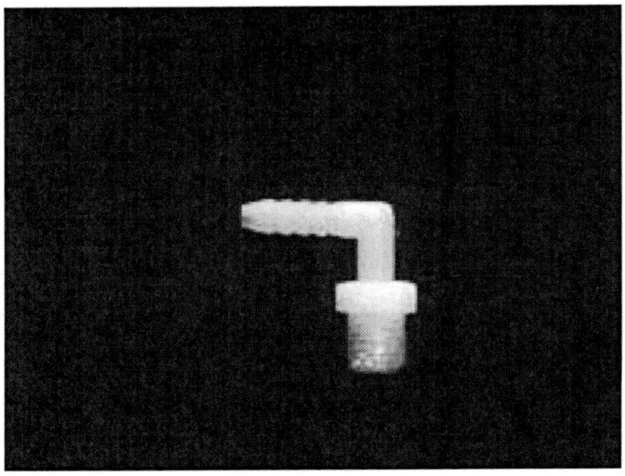

Figure: A plastic elbow fitting with a hose barb at the top, and male pipe threads at the bottom

A barb is used to connect flexible hoses to pipes. A barb clamp fitting has a male-threaded end used to mate with the female-threads. The other end of the fitting has either a single or multiple barbed tube having a tapered stub with ridges which is inserted into the flexible hose to secure it.

An adjustable worm driver screw clamp helps to keep the hose from slipping off the barbed tube. Barb fittings can be made of plastic or brass. Brass is used for hot water applications while plastic is used for cold. The barb fitting can be either elbow-shaped or straight.

Valves

Valve is equipment designed to stop or regulate flow of any fluid (liquid, gas, condensate, stem, slurry etc.) in its path. Valves are categorised depending on their applications like isolation, throttling and non-return. It is installed in the piping system based on its requirement.

Various type of valves are available depending upon the type of construction as follows:

- Gate valve - used for isolation only
- Plug valve - used for isolation only
- Globe valve - used for throttling
- Butterfly valve - used for isolation as well as throttling
- Check valve - used for preventing reverse flow (non-return)
- Diaphragm valve - used for isolation as well as throttling
- Ball valve - used for isolation only.

Drain-waste-vent (DWV) and Related Fittings

Because they operate at low pressure and rely on gravity to move fluids (and often entrained solids), drain-waste-vent systems use fittings designed to be as smooth as possible on their interior surfaces. The fittings may be "belled" or expanded slightly in diametre, or otherwise shaped to accommodate insertion of pipe or tubing, without forming a sharp interior ridge that might catch debris or accumulate buildup of material and cause clogging. The absence of interior snags also makes it much easier to "snake out" or "rod out" a clogged pipe using long flexible tools made for this purpose.

Underground piping systems for landscaping drainage, or disposal of stormwater or groundwater, similarly use gravity flow at low pressure, often with entrained solids. Piping fittings used for these systems bear a strong resemblance to DWV fittings, though often at a larger scale. When high peak flow volumes are involved, the design and construction of these systems are closely inter-related to sewer design.

Fittings for central vacuum systems are very similar to DWV fittings, though usually of thinner and lighter construction, since the weight of the materials conveyed through the system is much less. Vacuum system designs share with DWV designs a concern about eliminating internal ridges, burrs, sharp turns, or other obstructions to smooth flow that might cause build-up of material into pipe blockages.

Sweep Elbow

DWV elbows are usually long radius or sweep types, to reduce flow resistance and solids deposition when the direction of flow is changed. A well-designed system will often employ multiple 45° elbows in preference over 90° elbows (even sweep elbows), to reduce flow disruption as much as possible.

Central vacuum system inlet fittings are intentionally designed with a tighter radius of curvature than any other bends in the system. This is done to insure that if any vacuumed debris becomes stuck, it will jam right at the inlet, where it is easiest to discover and to remove.

Closet Flange

The closet flange is the drain pipe flange to which a water closet (toilet) is attached. It is a specialised type of flange connection designed to sit flush with the floor, allowing a standard toilet to be installed above it.

Clean-outs

Clean-outs are fittings with removable elements that allow access to drains without requiring removal of plumbing fixtures. They are used for allowing an auger or plumber's snake to clean out a plugged drain. Clean-outs should be placed in accessible locations at regular intervals throughout a drainage system, often including outside the building, because clean-out augers have limited length. The minimum requirement is typically at the end of each branch in piping, just ahead of each water closet, at the base of each vertical stack, and both inside and outside the building in the building main drain/sewer. Clean-outs normally have screw-on caps or screw-in plugs. Clean-outs are also known as rodding eyes from the eye-shaped cover plates often used on external versions.

Trap Primers

Trap primers regularly inject water into traps so that "water seals" are maintained, as necessary to keep sewer gases out of buildings. The trap primer must be installed in a readily available place for easy access for adjustments, replacement, and repair. Strictly speaking, a trap primer is a specialised valve, and it is usually connected to a clean water supply, in addition to a DWV system. Because of this dual connection, the design usually must be certified to resist accidental backflow of contaminated water.

Combo-Tee

A combination tee (combo tee) is a tee with a gradually curving centre connecting joint. It is used in drain systems to provide a smooth, gradually curving path to reduce the likelihood of clogs, and to ease pushing a plumber's snake through a drain system. The "combo" is a combination of a wye and a 1/8 bend or 45° elbow.

Sanitary Tee

A sanitary tee is a tee with a curved centre section designed to minimise the possibility of siphon action that could draw water out of a trap. The centre connection is generally connected to the pipe which leads to a trap (the trap arm).

Double Sanitary Tee (Sanitary Cross)

Similar to a cross. This fitting differs from a standard cross in that two of the ports have curved inlets. The fitting has been used in the past for connecting the drains of back-to-back fixtures (such as back-to-back bathroom sinks). Some current codes (including the

2006 UPC) prohibit the use of this fitting for that purpose, instead requiring a special "double fixture fitting".

Wye Fitting

A fitting with three openings, a wye is used to create branch lines. It is a type of waste fitting tee which has the side inlet pipe entering at a 45° angle, or an angle other than 90 degrees. A standard wye is a "Y" shaped fitting which allows one pipe to be joined to another at a 45 degree angle.

Wyes are similar to tees except that the branch line is angled to reduce friction and turbulence that could hamper the flow. The connection is typically at a 45-degree angle rather than a 90-degree angle. If a branch turns out further at the end to be perpendicular to the through line, the fitting becomes a "tee wye" (TY).

PVDF Corrosive Waste Piping Systems utilise wye fittings that feature a smooth inside diametre without any irregularities, for uninterrupted flow path. These fittings should be able to withstand acids, bases, and solvents. Flame-retardant systems withstand intermittent corrosive fluids up to high temperatures. Polypropylene and PVDF Double Wyes and Double Reducing Wyes come in various sizes and are largely used for this purpose.

A wye branch allows to split a branch line equally in two directions. The opening sizes can vary for different situations, for instance in situation where a large main line needs to be split into two smaller branches.

Low-priced wyes are often spot-welded together, whereas industrial wyes have a continuous weld at each seam. In long-distance pipeline applications, a true Wye fitting is also engineered for closed system instrumentation pigging configurations or wherever a smooth pipe branch is required.

Double-tapped Bushing

A double-tapped bushing is a fitting that has opposing threads on the inside diametre of the bushing.

Hydraulic Fittings

Hydraulic systems use extremely high fluid pressures to create useful work, such as in the hydraulic actuators for powered machinery such as bulldozers and backhoes. Therefore, hydraulic fittings are designed and rated for much greater pressures than those experienced in general piping systems, and they are generally not compatible for use in general plumbing. Fittings are designed and constructed to resist leakage and sudden explosive failure. More information on hydraulics and their specialised fittings can be found in the hydraulic machinery article.

Connection Types

Much of the work of installing a piping or plumbing system involves making leakproof, reliable connections. Depending on the technology used, basic skills may be required, or specialised skills and professional licensure may be required.

Fastener

A fastener is a hardware device that mechanically joins or affixes two or more objects together. The following are the type of fasteners commonly used:

- Stud bolt with nut
- Machine bolt with nut

Usually the stud bolts are used with full threading and with two heavy hexagonal nuts.

Threaded Pipe

A threaded pipe is a pipe with a screw thread at one or both ends for assembly. Steel pipe is often joined using threaded connections, where tapered threads are cut into the end of the tubing segment, sealant is applied in the form of thread sealing compound or thread seal tape (also known as PTFE or Teflon tape), and it is then threaded into a corresponding threaded fitting using a pipe wrench.

Threaded steel pipe is still widely used in many homes and businesses to convey natural gas or propane fuel, and is a popular choice in fire sprinkler systems due to its high heat resistance. Threaded brass pipe was once used in a similar fashion, and was considered superior to steel for carrying drinking water, but is now effectively obsolete.

Assembling threaded steel pipe takes some skill, plus careful planning to allow lengths of pipe to be screwed together in proper sequence. Most threaded pipe systems require occasional use of pipe union fittings to allow final assembly.

Solvent Welding

A solvent is applied to PVC, CPVC, ABS, or other plastic piping, to partially dissolve and fuse the adjacent surfaces of piping and fitting. Solvent welding is usually used with a sleeve-type joint, to connect pipe and fittings made of the same (or closely compatible) material.

Unlike regular welding of metals, solvent welding is relatively easy to perform, although care is still needed to produce reliable joints. Solvents typically used for plastics are usually toxic, may be carcinogenic, and may also be flammable, requiring adequate ventilation.

Soldering

To make a solder connection, a chemical flux is applied to the inner sleeve of a sleeve type joint, and the pipe is inserted. The joint is then heated using a propane torch or MAPP gas torch, solder is applied to the heated joint, and the melted solder is drawn into the joint by capillary action as the flux vapourises. Sweating is an alternate term sometimes used to describe soldering of pipe joints.

In situations where many connections must be made in a short period of time (such as plumbing of a new building), solder offers much quicker and much less expensive joinery than compression or flare fittings. A degree of skill is needed to rapidly make large numbers of reliable soldered joints. If flux residues are thoroughly cleaned, soldering can produce a long-lasting connection at low cost. However, the use of open flames for heating joints can present fire and health hazards to occupants of a building being worked on, and requires adequate ventilation.

Brazing

Brazing is a thermal joining process in which the two pieces of the base metal are joined when a molten brazing filler metal is allowed to be drawn into a capillary gap between them. It is used to join most metals and alloys commonly used in engineering. Brazing filler metals have very high melting points, but always below the melting point of the metals being joined. It is closely related to soldering, but uses harder materials and higher temperatures. Brazing can be used to join pipes, rods, metal pieces or any other shape as long as the pieces fit neatly against each other without leaving large gaps. It is also capable of joining tungsten carbide, ceramics and similar non-metallic materials. Successfully brazed joints are as strong as the parent metal pieces being joined and can withstand demanding service conditions i.e they are strong and ductile. Well brazed joints have smooth, neat fillets and they offer good electrical conductivity.

Welding

Welding of metals differs from soldering and brazing, in that the connection is made without adding a special low-melting-point material (e.g. solder) to complete a joint. Instead, the material of the pipe or tubing is itself partially melted in a carefully controlled manner, and the fitting and piping are directly fused together. Generally this requires that the piping and the fitting be made of the same (or closely compatible) material.

Pipe welding is a specialised skill, and is often performed by specially licensed workers who are tested periodically for the quality of their work. For ultra-critical applications, every joint is tested using non-destructive testing methods.

Properly welded and inspected joints are considered to be very reliable, robust, and long-lasting. Proper ventilation is essential to remove dangerous metallic fumes from welding operations. Because of the expensive skilled labour required, welded pipe joints are usually restricted to high-performance applications, such as in shipbuilding, chemical reactors, and nuclear reactors.

Compression Fittings

Figure: Compression connectors on an isolating valve. Red metal is a copper compression ring.

Compression fittings consist of a tapered concave conical seat, a hollow barrel-shaped compression ring, and a compression nut which is threaded onto the body of the fitting and tightened to make a leakproof connection. Fittings are typically made of brass or plastic, but stainless steel or other materials may be used.

Compression connections do not typically have the long life that sweat connections offer, but are advantageous in many cases because they are easy to make using basic tools. A disadvantage of compression connections is that they take longer to make than sweated joints, and sometimes require retightening over time to stop leaks.

Flare Fittings

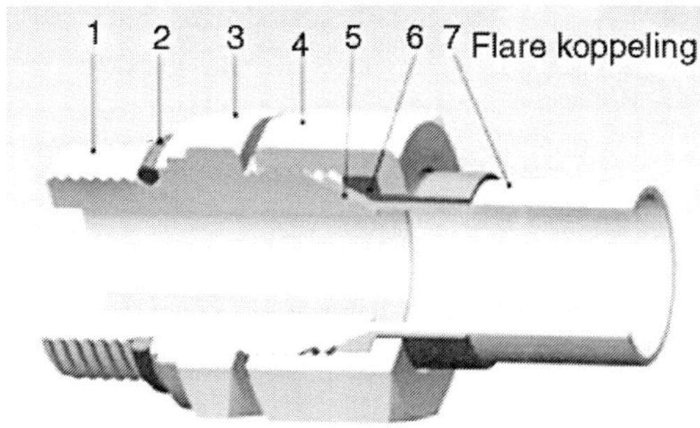

Figure: *Flare connection:* 1 Screw thread, 2 O-ring, 3 Body, 4 Nut, 5 Seal interface, 6 Support ring (sleeve), 7 Flared tubing.

Flared connectors should not be confused with compression connectors, with which they are generally not interchangeable. Flared connectors lack a compression ring, but do use a threaded nut. A special flaring tool is used to enlarge tubing into a tapered "bellmouth" shape that matches the tapered projecting conical shape of the flare fitting. The flare nut, which has previously been installed over the tubing, is then tightened onto the fitting. Fittings are typically made of brass or plastic, but stainless steel or other materials may be used.

Flare connections are a labour intensive method of making connections, but are quite reliable over the course of many years. Flared fittings are sometimes thought to be more secure against leaks and sudden failures, and are often preferred for safety-critical connections, such as in hydraulic brake systems.

Flange Fittings

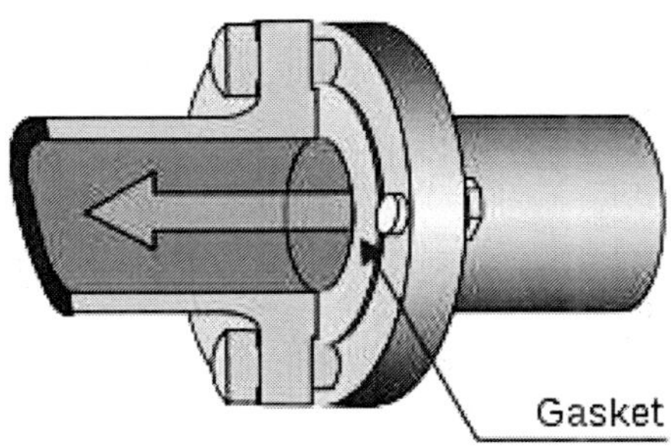

Figure: Flange connection, using a gasket

Flanges are generally used when there is a connection to valves, in-line instruments and/or connection to equipment nozzles is required.

Flange fittings generally involve pressing two surfaces to be joined tightly together, by means of threaded bolts, wedges, clamps, or other means of applying high compressive forces. Often, a gasket, packing, or an O-ring is installed between the flanges to prevent leakage, but it is sometimes possible to use only a special grease, or nothing at all, if the mating surfaces are precisely formed. Flanges are designed to the following pressure ratings: 150 lb, 300 lb, 400 lb, 600 lb, 900 lb,1500 lb and 2500 lb or 10 Bar, 15Bar, 25Bar, 40Bar, 64Bar, 100Bar and 150Bar. Various types of flanges are available depending upon the type of their constructional features. The following are types of flanges generally used in piping. These flanges are available with different facing like raised face, flat face, ring joint face etc.

- Orifice
- Threaded
- Slip-On
- Blind
- Weld Neck
- Socket
- Lap Joint
- Reducing.

Slip On flanges are slipped over the pipe and then welded from both inside and outside to provide sufficient strength and prevent leakage. This flange is used instead of weld necks by many users because of its lower cost and also the fact that it requires less accuracy when cutting pipe to length.

Blind flanges do not have a bore and are used to shut off a piping system or vassal opening. Its design permits easy access to vassal or piping system for inspection purpose. It can be supplied with or without hubs at the manufacturer's option.

Weld Neck Flanges are designed to be joined to a piping system by butt welding. They are expensive because of its long neck, but is preferred for high stresses to the pipe, reducing stress applications. The neck, or hub, transmits stress concentration at the base of the flange. The gradual transition of thickness from the base of the hub to the wall thickness at the butt weld provides important reinforcement of the flange. Turbulence and erosion are reduced due to the matching bore size of the pipe and flange.

Socket Flanges are similar to a slip on flanges in outline, but the bore is counter-bored to accept pipe. The diametre of the remaining bore is same as the inside diametre of the pipe. A fillet weld around the hub of the flange attaches the flange to the pipe. An optional interval weld may be applied in high stress applications. Its biggest use is in high pressure system such as hydraulic and steam lines.

Lap Joint flange is again similar to a slip flange, but it has radius at the intersection of the bore and the flange face to accommodate a lap stub end. The face on the stub end forms the gasket face of the flange. Its applications are where sections of piping systems need to be dismantled quickly and easily for inspection or replacement.

Flanged connections tend to be more bulky than other connections, but can perform well in demanding applications, such as large water mains and hydroelectric power systems.

Mechanical Fittings

Figure: Multiple mechanical sleeve clamps connecting piping

Manufacturers such as Victaulic or Grinnell produce special sleeve clamp fittings that are increasingly replacing classic flanged connections. They typically attach to the end of a pipe segment by using circumferential grooves pressed (or cut, in older designs) around the end of the pipe to be joined. Mechanical connectors are widely used on larger steel pipes, but can also be used with other materials.

The chief advantage of these newer connectors is that they can be installed in the field after cutting the pipe to length, which is much faster than traditional flanged connections, which must be factory-welded or field-welded to pipe segments.

Crimped or Pressed Fittings

Crimped or pressed connections use special fittings which are permanently attached to tubing with a powered crimper. The special fittings, manufactured with sealant already inside, slide over the tubing to be connected. High pressure is used to deform the fitting and compress the sealant against the inner tubing, creating a leakproof seal. The advantages of this method are that it should last as long as the tubing, it takes less time to complete than other methods, it is cleaner in both appearance and the materials used to make the connection, and no open flame is used during the connection process. The disadvantages are that the fittings used are harder to find and cost significantly more than sweat type fittings.

Fittings and Valves

Fittings are also used to split or join a number of pipes together, and for other purposes. A broad variety of standardised pipe fittings are available; they are generally broken down into either a tee, an elbow, a branch, a reducer/enlarger, or a wye. Valves control fluid flow and regulate pressure. The piping and plumbing fittings and valves articles discuss them further.

Tube Cleaning

Tube cleaning describes the activity of, or device for, the cleaning and maintenance of fouled tubes. The need for cleaning arises because the medium that is transported through the tubes may cause deposits and finally even obstructions. In system engineering and in industry, particular demands are placed upon surface roughness or heat transfer. In the food and pharmaceutical industries as well as in medical technology, the requirements are germproofness, and that the tubes are free from foreign matter, for example after the installation of the tube or after a change of product. Another trouble source may be corrosion due to deposits which may also cause tube failure. Depending on application, conveying medium and tube material, the following methods of tube cleaning are available:

Lost Tubes

In medical field, "lost" tubes are tubes which have to be replaced after single use. This is not genuine tube cleaning in the proper sense and is very often applied in the medical sector, for instance with cannulas of syringes, infusion needles or medical appliances, such as kidney machines at dialysis. The reasons for the single use are primarily the elimination of infection risks but also the fact that cleaning would be very expensive and, particularly with cheap mass products, out of all proportion in terms of cost. Single use is therefore common practice with tubes of up to 20 mm diametre. For the same reasons as in the medical sector, single use may also be applicable in the food and pharmaceutical process technology, however in these sectors the tube diametres may exceed 20 mm. In other fields (e.g., in heat exchangers), tubing may also sometimes need to be replaced (or removed, plugged, etc), but this typically occurs only after a prolonged use, when the tube develops serious flaws (e.g., due to corrosion).

Chemical Tube Cleaning

Chemical tube cleaning is understood to be the use of cleaning liquids or chemicals for removing layers and deposits. A typical example is the deliming of a coffee maker where scale is removed by means of acetic acid or acid of lemon. Depending on the field of application and tube material, special cleaning liquids may be used which also require a multi-stage treatment:

- chemical activation
- cleaning
- rinsing

This method of cleaning calls for a shutdown of the relevant system which causes undesired standstill periods. To safeguard a continuous production operation it may be necessary to install several systems. Another disadvantage: in the field of large-scale technology (reactor, heat exchanger, condenser, etc.), huge quantities of cleaning liquids would be required which would cause major disposal problems. A further problem occurs in the food industry through the possible toxicity of the cleaning liquid. Only the strict observance of rinsing instructions and an exact control of the admissible residue tolerances can remedy things here. This in turn requires expensive detection methods. Generally the process of chemical tube cleaning is applicable for any diametre, however practical limits of use ensue from the volume of a pipeline.

Mechanical Tube Cleaning

A mechanical tube cleaning system is a cleaning body that is moved through the tube in order to remove deposits from the tube wall. In the most simple case it is a matter of a brush that is moved in the tube by means of a rod or a flexible spring (device). In large-scale technology and industrial sector, however, several processes have developed which necessitate a more detailed definition.

Off-line Process

An off-line process is characterised by the fact that the system to be cleaned has to be taken out of operation in order to inject the cleaning body(ies) and to execute the cleaning procedure. An additional distinction must be made between active and passive cleaning bodies.

Passive cleaning bodies may be a matter of brushes or special constructions like scrapers or so-called "pigs", for instance, which are conveyed through the tubes by means of pressurised air, water, or other media. In most cases, cleaning is implemented through the oversize of the cleaning bodies compared to the tube inner diametre. The types range from brushes with bristles of plastic or steel to scrapers (with smaller tube diametres) and more expensive designs with spraying nozzles for pipelines. This method is applied for tube and pipe diametres from around 5 mm to several metres. Also belonging to this field is the cleaning of obstructed soil pipes of domestic sewage systems that is done by means of a rotating, flexible shaft.

The active cleaning bodies are more or less remote controlled robots that move through the tubes and fulfill their cleaning task, pulling along with them not only cables for power supply and communication but also hoses for the cleaning liquid. Also measuring devices or cameras are carried along to monitor the function. To date, such devices have still required minimum diametres of around 300 mm, however a further diminution is being worked on. The reasonable maximum diametre of this kind of devices is 2 m because above this diametre an inspection of the pipe would certainly be less expensive. For such large diametres a robot application is imaginable only if health-hazardous chemicals are in use.

On-line Process

In the on-line process, the cleaning body moves through the tubes with the conveying medium and cleans them by means of its oversize compared to the tube diametre. In the range of diametres of up to 50 mm these cleaning bodies consist of sponge rubber, in larger diametres up to the size of oil pipelines it is a matter of scrapers or so-called pigs. Sponge rubber balls are applied mainly for cooling water, like sea, river, or cooling tower water. For the chemical or pharmaceutical industry, specially adapted cleaning bodies are imaginable but the conveying media flows are so weak that off-line processes are employed in most cases. Given the fact that the cleaning bodies are not allowed to remain in the conveying medium they have to be collected after passing through the tubes. In the case of sponge rubber balls this is done through special strainer sections; for scrapers or pigs an outward transfer station is provided. According to the Taprogge process, the sponge rubber balls are re-injected upstream of the system to be cleaned by a corresponding ball recirculating unit whereas the scraper or pig is mostly taken out by hand and re-injected into another collector. Sponge rubber balls therefore safeguard a continuous cleaning while the scraper or pig system is discontinuous.

Thermal Tube Cleaning

In thermal tube cleaning the layer or deposit is dried through a heating whereby it flakes off due to its embrittlement and is then discharged, either by the conveying medium or a rinsing liquid. Depending on the required temperature, the heating can be either a parallel tube heating or an induction heating. This process is an off-line process. Occasionally it is also used for the sterilization of tubes in the pharmaceutical or food industry. A diametre range cannot be indicated here because this method can be applied for certain processes only; a technical limitation of the heating is given only by the materials and the required amount of heat.

Special Types

Special types of tube cleaning are all such types which are partly in experimental stage only and do not come under the process types mentioned before, such as, for example:

- induction of water hammers, so that the layer or deposit comes off through short-term material elongation
- use of vibration generators, partly at the tubes through vibration exciters, partly by means of piezoelectric crystals in the conveying medium, in order to transform the conveying medium into a cleaning medium through reduction of the surface tension
- magnetic fields to avoid tube calcification
- nanotechnical treatment of tube surfaces to avoid layers and deposits.

Piping

Within industry, piping is a system of pipes used to convey fluids (liquids and gases) from one location to another. The engineering discipline of piping design studies the efficient transport of fluid.

Industrial process piping (and accompanying in-line components) can be manufactured from wood, fibreglass, glass, steel, aluminium, plastic, copper, and concrete. The in-line components, known as fittings, valves, and other devices, typically sense and control the pressure, flow rate and temperature of the transmitted fluid, and usually are included in the field of Piping Design (or Piping Engineering). Piping systems are documented in piping and instrumentation diagrams (P&IDs). If necessary, pipes can be cleaned by the tube cleaning process.

Figure: Mechanical equipment and electrical piping in the mechanical room

"Piping" sometimes refers to Piping Design, the detailed specification of the physical piping layout within a process plant or commercial building. In earlier days, this was sometimes called Drafting, Technical drawing, Engineering Drawing, and Design but is today commonly performed by Designers who have learned to use automated Computer Aided Drawing / Computer Aided Design (CAD) software.

Plumbing is a piping system that most people are familiar with, as it constitutes the form of fluid transportation that is used to provide potable water and fuels to their homes and business. Plumbing pipes also remove waste in the form of sewage, and allow venting of sewage gases to the outdoors. Fire sprinkler systems also use piping, and may transport nonpotable or potable water, or other fire-suppression fluids.

Piping also has many other industrial applications, which are crucial for moving raw and semi-processed fluids for refining into more useful products. Some of the more exotic materials of construction are Inconel, titanium, chrome-moly and various other steel alloys.

Engineering Subfields

Generally, Industrial Piping Engineering has three major subfields:

- Piping Material
- Piping Design
- Stress Analysis.

Stress Analysis

Process piping and power piping are typically checked by Pipe Stress Engineers to verify that the routing, nozzle loads, hangers, and supports are properly placed and selected such that allowable pipe stress is not exceeded under different situation such as sustain, operating, pressure testing etc, as per the ASME B31, EN 13480 or any other applicable codes and standards. It is necessary to evaluate the mechanical beahviour of the piping under regular loads (internal pressure and thermal stresses) as well under occasional and intermittent loading cases such as earthquake, high wind or special vibration, and water hammer. This evaluation is usually performed with the assistance of a specialised (finite element) pipe stress analysis computer program such as CAESAR II, ROHR2, CAEPIPE and AUTOPIPE.

In cryogenic pipe supports, most steel become more brittle as the temperature decreases from normal operating conditions, so it is necessary to know the temperature distribution for cryogenic conditions. Steel structures will have areas of high stress that may be caused by sharp corners in the design, or inclusions in the material.

Wooden Piping History

Early wooden pipes were constructed out of logs that had a large hole bored lengthwise through the centre. Later wooden pipes were constructed with staves and hoops similar to wooden barrel construction. Stave pipes have the advantage that they are easily transported as a compact pile of parts on a wagon and then assembled as a hollow structure at the job site. Wooden pipes were especially popular in mountain regions where transport of heavy iron or concrete pipes would have been difficult.

Wooden pipes were easier to maintain than metal, because the wood did not expand or contract with temperature changes as much as metal and so consequently expansion joints and bends were not required. The thickness of wood afforded some insulating properties to the pipes which helped prevent freezing as compared to metal pipes. Wood used for water pipes also does not rot very easily. Electrolysis, that bugbear of many iron pipe systems, doesn't affect wood pipes at all, since wood is a much better electrical insulator.

In the Western United States where redwood was used for pipe construction, it was found that redwood had "peculiar properties" that protected it from weathering, acids, insects, and fungus growths. Redwood pipes stayed smooth and clean indefinitely while iron pipe by comparison would rapidly begin to scale and corrode and could eventually plug itself up with the corrosion.

Materials

The material with which a pipe is manufactured often forms as the basis for choosing any pipe. Materials that are used for manufacturing pipes include:

- Carbon Steel (CS)
- Low Temperature Service Carbon Steel (LTCS)
- Stainless Steel (SS)
- Non Ferrous Metals (Inconel, Incoloy, Cupro-nickel, etc.)
- Non Metallic (GRE, PVC, HDPE, tempered glass, etc.)

Standards

There are certain standard codes that need to be followed while designing or manufacturing any piping system. Organisations that promulgate piping standards include:

- ASME - The American Society of Mechanical Engineers - B31 series
  - ASME B31.1 Power piping (steam piping etc)
  - ASME B31.3 Process piping
  - ASME B31.4 Pipeline Transportation Systems for Liquid Hydrocarbons and Other Liquids
  - ASME B31.5 Refrigeration piping and heat transfer components
  - ASME B31.8 Gas transmission and distribution piping systems
  - ASME B31.9 Building services piping
  - ASME B31.11 Slurry Transportation Piping Systems

  - ASME B31.12 Hydrogen Piping and Pipelines
- ASTM - American Society for Testing and Materials
- API - American Petroleum Institute
- EN 13480 - European metallic industrial piping code
  - EN 13480-1 Metallic industrial piping - Part 1: General
  - EN 13480-2 Metallic industrial piping - Part 2: Materials
  - EN 13480-3 Metallic industrial piping - Part 3: Design and calculation
  - EN 13480-4 Metallic industrial piping - Part 4: Fabrication and installation
  - EN 13480-5 Metallic industrial piping - Part 5: Inspection and testing
  - EN 13480-6 Metallic industrial piping - Part 6: Additional requirements for buried piping
  - PD TR 13480-7 Metallic industrial piping - Part 7: Guidance on the use of conformity assessment procedures
  - EN 13480-8 Metallic industrial piping - Part 8: Additional requirements for aluminium and aluminium alloy piping
- EN 1993-4-3 Eurocode 3 — Design of steel structures - Part 4-3: Pipelines
- AWS - American Welding Society
- AWWA - American Water Works Association
- MSS – Manufacturers' Standardisation Society
- ANSI - American National Standards Institute
- NFPA - National Fire Protection Association
- EJMA - Expansion Joint Manufacturers Association.

Chapter 2

Pipe Network Analysis

In fluid dynamics, pipe network analysis is the analysis of the fluid flow through a hydraulics network, containing several or many interconnected branches. The aim is to determine the flow rates and pressure drops in the individual sections of the network. This is a common problem in hydraulic design.

Description

In order to direct water to many individuals in a municipal water supply, many times the water is routed through a water supply network. A major part of this network may consist of interconnected pipes. This network creates a special class of problems in hydraulic design typically referred to as pipe network analysis. The modern solution for this is to use specialised software in order to automatically solve the problems. However, the problems can also be addressed with simpler methods like a spreadsheet equipped with a solver, or a modern graphing calculator.

Network Analysis

Once the friction factors are solved for, then we can start considering the network problem. We can solve the network by satisfying two conditions.

1. At any junction, the flow into a junction equals the flow out of the junction.
2. Between any two junctions, the head loss is independent of the path taken.

The classical approach for solving these networks is to use the Hardy Cross method. In this formulation, first you go through and create guess values for the flows in the network. That is, if Q7 enters a junction and Q6 and Q4 leave the same junction, then the initial guess must satisfy $Q7 = Q6 + Q4$. After the initial guess is made, then, a loop is considered so that we can evaluate our second condition.

Given a starting node, we work our way around the loop in a clockwise fashion, as illustrated by Loop 1.

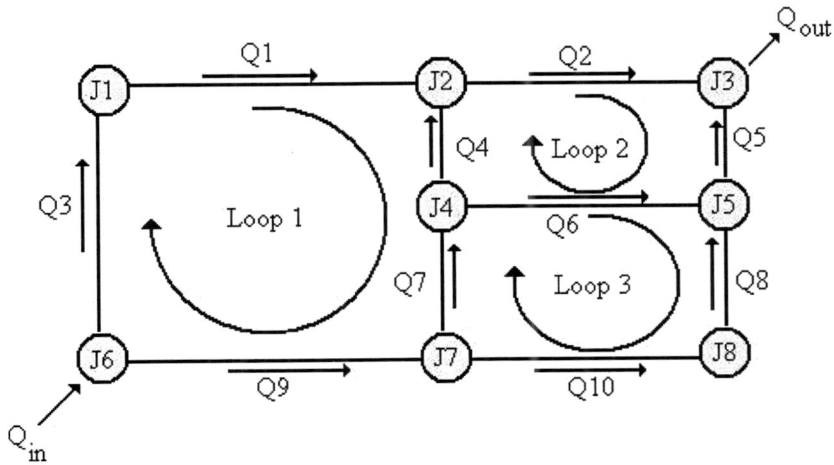

We add up the head losses according to the Darcy–Weisbach equation for each pipe if Q is in the same direction as our loop like Q1, and subtract the head loss if the flow is in the reverse direction, like Q4. In order to satisfy the second condition, we should end up with 0 about the loop if the network is completely solved.

If the actual sum of our head loss is not equal to 0, then we will adjust all the flows in the loop by an amount given by the following formula, where a positive adjustment is in the clockwise direction.

$$\Delta Q = \frac{\sum \text{head loss}_c - \sum \text{head loss}_{cc}}{n \cdot (\sum \frac{\text{head loss}_c}{Q_c} + \sum \frac{\text{head loss}_{cc}}{Q_{cc}})},$$

where

- n is 1.85 for Hazen-Williams and
- n is 2 for Darcy–Weisbach.

The clockwise specifier (c) means only the flows that are moving clockwise in our loop, while the counter-clockwise specifier (cc) is only the flows that are moving counter-clockwise.

This adjustment won't solve the problem, since with most networks we will have several loops. It is ok to do this adjustment, however, because our flow changes won't alter condition 1, and therefore, our other loops will still satisfy condition 1. However, we should use the results from the first loop if we progress to any other loops.

The more modern method is simply to create a set of conditions from your junctions and head-loss criteria. Then, use a Root-finding algorithm to find Q values that satisfy all the equations. The literal friction loss equations will use a term called Q^2, but we want to preserve any changes in direction. Create a separate equation for each loop where the head losses are added up, but instead of squaring Q, use $|Q| \cdot Q$ instead (with $|Q|$ the absolute value of Q) for the formulation so that any sign changes will reflect appropriately in the resulting head-loss calculation.

Gas Networks Simulation

Gas networks simulation is a process of defining the mathematical model of the gas transport and distribution systems, which are usually composed of highly integrated pipe networks operating over a wide range of pressures. Simulation allows to predict the behaviour of gas network systems under different conditions. Such predictions can be effectively used to guide decisions regarding the design and operation of the real system.

Simulation Types

Depending on the gas flow characteristics in the system there are two states that can be matter of simulation:

- Steady state - the simulation does not take into account the gas flow characteristics' variations over time and described by the system of algebraic equations, in general nonlinear ones.
- Unsteady state (transient flow analysis) - described either by a partial differential equation or a system of such equations. Gas flow characteristics are mainly functions of time.

Network Topology

In the gas networks simulation and analysis, matrices turned out to be the natural way of expressing the problem. Any network can be described by set of matrices based on the network topology. Consider the gas network by the graph below. The network consists of one source node (reference node) L1, four load nodes (2, 3, 4 and 5) and seven pipes or branches. For network analysis it is necessary to select at least one reference node. Mathematically, the reference node is referred to as the independent node and all nodal and branch quantities are dependent on it. The pressure at source node is usually known, and this node is often used as the reference node. However, any node in the network may have its pressure defined and can be used as the reference node. A network may contain several sources or other pressure-defined nodes and these form a set of reference nodes for the network.

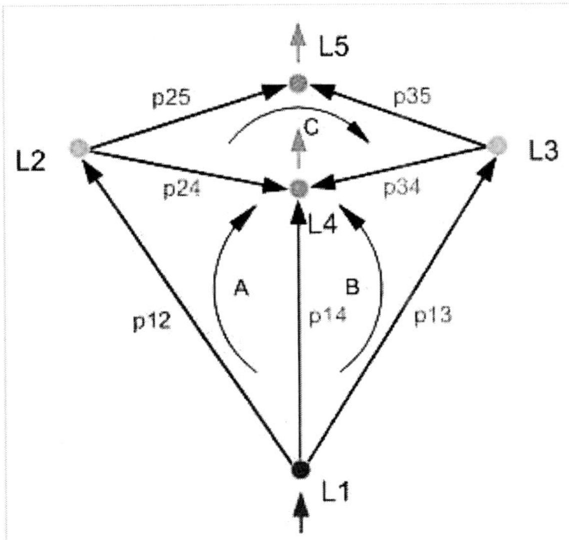

The load nodes are points in the network where load values are known. These loads may be positive, negative or zero. A negative load represents a demand for gas from the network. This may consist in supplying domestic or commercial consumers, filling gas storage holders, or even accounting for leakage in the network. A positive load represents a supply of gas to the network. This may consist in taking gas from storage, source or from another network. A zero load is placed on nodes that do not have a load but are used to represent a point of change in the network topology, such as the junction of several branches. For steady state conditions, the total load on the network is balanced by the inflow into the network at the source node.

The interconnection of a network can produce a closed path of branches, known as a loop. A consists of branches p12-p24-p14, loop B consists of p13-p34-p14, and loop C consists of p24-p25-p35-p34. A fourth loop may be defined as p12-p24-p34-p13, but it is redundant if loops A, B and C are also defined. Loops A, B and C are independent ones but the fourth one is not, as it can be derived from A, B and C by eliminating common branches. To define the network topology completely it is necessary to assign a direction to each branch. Each branch direction is assigned arbitrarily and is assumed to be positive direction of flow in the branch. If the flow has the negative value, then the direction of flow is opposite to branch direction. In the similar way, direction is assigned to each loop and flow in the loop.

The solutions of problems involving gas network computation of any topology requires such a representation of the network to be found which enables the calculations to be performed in the most simple way. These requirements are met by the graph theory which permits representation of the network structure by means of the incidence properties of the network components and, in consequence, makes such a representation explicit.

Flow Equations

The calculation of the pressure drop along the individual pipes of a gas network requires use of the flow equations. Many gas flow equations have been developed and a number have been used by the gas industry. Most are based on the result of gas flow experiments. The result of the particular formula normally varies because these experiments were conducted over different range of flow conditions, and on varying internal surface roughness. Instead, each formula is applicable to a limited range of flow and pipe surface conditions.

Mathematical Methods of Simulation

Steady State Analysis

A gas network is in the steady state when the values of gas flow characteristics are independent of time and system described by the set of nonlinear equations. The goal of simple simulation of a gas network is usually that of computing the values of nodes' pressures, loads and the values of flows in the individual pipes. The pressures at the nodes and the flow rates in the pipes must satisfy the flow equations, and together with nodes' loads must fulfill the first and second Kirchhoff's laws. There are many methods of analysing the mathematical models of gas networks but they can be divided into two types as the networks, the solvers for low pressure networks and solvers for high pressure networks.

The networks equations are nonlinear and are generally solved by some of Newton iteration; rather than use the full set of variables it is possible to eliminate some of them. Based on the type of elimination we can get solution techniques are termed either nodal or loop methods.

Newton-nodal Method

The method is based on the set of the nodal equations which are simply mathematical representation of Kirchhof's first law which states that the inlet and outlet flow at each node should be equal. Initial approximation is made to the nodal pressures.

The approximation is then successively corrected until the final solution is reached.

Disadvantages

- Poor convergence characteristics, the method is extremely sensitive to initial conditions.

Advantages

- Does not require extra computation to produce and optimise a set of loops.
- Can easily adapted for optimisation tasks.

Newton-loop Method

The method is based on the generated loops and the equations are simply mathematical representation of Kirchhof's second law which states that the sum of the pressure-drops around any loop should be zero. Before using loops method the fundamental set of loops need to be found. Basically the fundamental set of loops can be found by constructing spanning tree for the network. The standard methods for producing spanning tree is based on a breadth-first search or on a depth-first search which are not so efficient for large networks, because the computing time of these methods is proportional to n^2, where n is the number of pipes in the network. More efficient method for large networks is the forest method and its computational time is proportional to $n^*\log_2 n$.

The loops that are produced from the spanning tree are not the best set that could be produced. There is often significant overlap between loops with some pipes shared between several loops. This usually slows convergence, therefore the loops' reduction algorithm needs to be applied to minimise the loops overlapping. This is usually performed by replacing the loops in the original fundamental set by smaller loops produced by linear combination of the original set.

Disadvantages

- It requires extra computation to produce and optimise a set of loops.
- The dimension of the equations to be solved is smaller but they are much less sparse.

Advantages

- The main advantage is that the equation can be solved very efficiently with an iterative method that avoids the need of matrix factorisation and consequently has a minimal requirement for storage; this makes it very attractive for low pressure networks with a large number of pipes.

- Fast convergence which is less sensitive to the initial conditions.

Newton loop-node Method

The Newton loop-node method is based on Kirchhoff's first and second laws. The Newton loop-node method is the combination of the Newton nodal and loop methods and does not solve loop equations explicitly. The loop equations are transformed to an equivalent set of nodal equations, which are then solved to yield the nodal pressures. The nodal pressures are used then to calculate the corrections to the chord flows (which is synonymous to loop flows), and the tree branch flows are obtained from them.

Disadvantages

- Since set of nodal equations are solved nodal Jacobi matrix is used which is more sparse then the equivalent loop Jacobi matrix which may have negative impact on computational efficiency.

Advantages

- Good convergence characteristics of loop method are maintained.
- No need to define and optimise the loops.

Unsteady State Analysis

Computer Simulation: The importance of the mathematical methods' efficiency arises from the large scale of simulated network. It is required that the computation costs of the simulation method be low, this is related to the computation time and computer storage. At the same time the accuracy of the computed values must acceptable for the particular model.

Plastic Pipework

Plastic pipework is used for the conveyance of drinking water, waste water, chemicals, heating and cooling fluids, foodstuffs, ultra-pure liquids, slurries, gases, compressed air and vacuum system applications.

Materials Used

- ABS (Acrylonitrile Butadiene Styrene)
- UPVC (Unplasticized Polyvinyl Chloride)

- CPVC (Post Chlorinated Polyvinyl Chloride)
- PP (Polypropylene)
- PE (Polyethylene) also known as LDPE, MDPE and HDPE (low, medium, and high density)
- PVDF (Polyvinylidene fluoride).

Material Characteristics

ABS (Acrylonitrile Butadiene Styrene): ABS is used for the conveyance of potable water, slurries and chemicals.

UPVC (Unplasticized Polyvinyl Chloride) and CPVC (Post Chlorinated Polyvinyl Chloride)

UPVC has excellent chemical resistance across its operating temperature range, with a broad band of operating pressures. Due to its long-term strength characteristics, high stiffness and cost effectiveness, UPVC systems account for a large proportion of plastic piping installations. CPVC is resistant to many acids, bases, salts, paraffinic hydrocarbons, halogens and alcohols. It is not resistant to solvents, aromatics and some chlorinated hydrocarbons.

PP (Polypropylene)

Polypropylene is suitable for use with foodstuffs, potable and ultra pure waters, as well as within the pharmaceutical and chemical industries.

PE (Polyethylene)

Polyethylene has been successfully used for the safe conveyance of potable and waste water, hazardous waste, and compressed gases for many years.

PVDF (Polyvinylidene Fluoride)

PVDF has excellent chemical resistance which means that it is widely used in the chemical industry as a piping system for aggressive liquids.

Components of Plastic Pressure Pipe Systems

Pipes, fittings, valves, and accessories make up a plastic pressure pipe system. The range of pipe diametres for each pipe system does vary. However, the size ranges from 12 to 400 mm (0.472 to 15.748 in) and $^{3}D_{8}$ to 16 in (9.53 to 406.40 mm).

Pipes are extruded and are generally available in: 3 m (9.84 ft),4 m (13.12 ft), 5 m (16.40 ft), and 6 m (19.69 ft) straight lengths and 25 m

(82.02 ft), 50 m (164.04 ft), 100 m (328.08 ft), and 200 m (656.17 ft) coils for LDPE and MDPE.

Pipe fittings are moulded and come in many sizes: Tee 90° equal (straight and reducing), Tee 45°, Cross equal, Elbow 90° (straight and reducing), Elbow 45°, Short Radius Bend 90° Socket/coupler (straight and reducing), Union, End Caps, Reducing Bush, and Stub, Full face, and blanking flanges.

Valves are moulded and also come in many types: Ball Valve (also multiport), Butterfly valve, spring-, ball-, and swing-check non-return valves, Diaphragm valve, Knife Gate Valve, Globe Valve and Pressure Relief/Reduction Valves.

Reinforced Thermoplastic Pipe

RTP is a generic term referring to a reliable high strength synthetic fibre (such as Glass, Aramid or Carbon) reinforced thermoplastic pipe, initially developed in the early 1990s by Wavin Repox, Akzo Nobel and by Tubes d'Aquitaine from France, who developed the first pipes reinforced with synthetic fibre to replace medium pressure steel pipes in response to growing demand for non-corrisive conduits for application in the onshore oil and gas industry, particularly in the Middle East.

Typically, the materials used in the construction of the pipe might be Polyethylene (PE), Polyamide-11 or PVDF and may be reinforced with Aramid or Polyester fibre although other combinations are used.

More recently the technology of producing such pipe, including the marketing, rests with a few key companies, one of which is Pipelife with Soluforce where it is available in coils up to 400 m (1,312 ft) length. These pipes are available in pressure ratings from 30 to 90 bar (3 to 9 MPa; 435 to 1,305 psi). Over the last few years this type of pipe has been acknowledged as a standard alternative solution to steel for oilfield flowline applications by certain oil companies and operators. The great advantage of this pipe is also its very fast installation time compared to steel pipe when considering the welding time as average speeds up to 1,000 m (3,281 ft)/day have been reached installing RTP in ground surface.

Primarily, the pipe provides benefit to applications where steel may rupture due to corrosion and installation time is an issue.

Technology and History

The idea of synthetic fibre reinforced pipe has origins in the flexible hose and offshore industry where it has been frequently used for applications such as control lines in umbilicals and production flowlines for over 30 years. However, the commercialisation and realisation of a competitive product for the onshore oil industry came from a partnership between Akzo Nobel (supplier of Aramid fibre) and Wavin Repox (manufacturer of reinforced thermoset pipes), where Bert Dalmolen initiated a project to develop such a pipe. He was later employed by Pipelife where a state of the art production line was developed to produce RTP.

Pipelife also developed a pipe reinforced with steel wire to achieve even higher pressure ratings of over 150 bar (15 MPa; 2,176 psi) using steel reinforcement. Mr Chevrier (Tubes d'Aquitaine) also developed machinery that could produce such pipes, but was not successful in commercialising RTP.

Pipeline Transport

Pipeline transport is the transportation of goods through a pipe. Most commonly, liquids and gases are sent, but pneumatic tubes that transport solid capsules using compressed air are also used.

As for gases and liquids, any chemically stable substance can be sent through a pipeline. Therefore sewage, slurry, water, or even beer pipelines exist; but arguably the most valuable are those transporting fuels: oil (oleoduct), natural gas (gas grid), and biofuels.

Dmitri Mendeleev first suggested using a pipe for transporting petroleum in 1863.

Types by Transported Substance

For Oil or Natural Gas

Figure: A "Pig" launcher/receiver, belonging to the natural gas pipeline in Switzerland.

There is some argument as to when the first crude oil pipeline was built. However, some say pipeline transport was pioneered by Vladimir Shukhov and the Branobel company in the late 19th century. Others say oil pipelines originated when the Oil Transport Association first constructed a 2-inch (51 mm) wrought iron pipeline over a 6-mile (9.7 km) track from an oil field in Pennsylvania to a railroad station in Oil Creek, in the 1860s. Pipelines are generally the most economical way to transport large quantities of oil, refined oil products or natural gas over land. Compared to shipping by railroad, they have lower cost per unit and higher capacity. Although pipelines can be built under the sea, that process is economically and technically demanding, so the majority of oil at sea is transported by tanker ships.

The market size for oil and gas pipeline construction experienced tremendous growth prior to the economic downturn in 2008. The industry grew from $23 billion in 2006 to $39 billion in 2008.

Oil pipelines are made from steel or plastic tubes with inner diametre typically from 4 to 48 inches (100 to 1,200 mm). Most pipelines are typically buried at a depth of about 3 to 6 feet (0.91 to 1.8 m). The oil is kept in motion by pump stations along the pipeline, and usually flows at speed of about 1 to 6 metres per second (3.3 to 20 ft/s). Multi-product pipelines are used to transport two or more different products in sequence in the same pipeline. Usually in multi-product pipelines there is no physical separation between the different products. Some mixing of adjacent products occurs, producing interface. At the receiving facilities this interface is usually absorbed in one of the product based on pre-calculated absorption rates.

Crude oil contains varying amounts of wax, or paraffin, and in colder climates wax buildup may occur within a pipeline. Often these pipelines are inspected and cleaned using pipeline inspection gauges, *pigs*, also known as *scrapers* or *Go-devils*. Smart pigs (also known as intelligent or intelligence pigs) are used to detect anomalies in the pipe such as dents, metal loss caused by corrosion, cracking or other mechanical damage. These devices are launched from pig-launcher stations and travel through the pipeline to be received at any other station down-stream, either cleaning wax deposits and material that may have accumulated inside the line or inspecting and recording the condition of the line. For natural gas, pipelines are constructed of carbon steel and vary in size from 2 to 60 inches (51 to 1,500 mm) in diametre, depending on the type of pipeline. The gas is pressurised by compressor stations and is odourless unless mixed with a mercaptan odorant where required by a regulating authority.

For Ammonia

Highly-toxic ammonia is theoretically the most dangerous substance to be transported through long-distance pipelines. However, practically incidents on ammonia-transporting lines are uncommon-unlike on industrial ammonia-processing equipment.

Figure: The world's longest ammonia pipeline from Russia to Ukraine.

A major example of ammonia pipeline is the Ukrainian *Transammiak* line connecting the TogliattiAzot facility in Russia to the exporting Black Sea-port of Odessa.

For Biofuels (Ethanol and Biobutanol)

Pipelines have been used for transportation of ethanol in Brazil, and there are several ethanol pipeline projects in Brazil and the United States. Main problems related to the shipment of ethanol by pipeline are its high oxygen content, which makes it corrosive, and absorption of water and impurities in pipelines, which is not a problem with oil and natural gas. Insufficient volumes and cost-effectiveness are other considerations limiting construction of ethanol pipelines.

For Coal and Ore

Slurry pipelines are sometimes used to transport coal or ore from mines. The material to be transported is closely mixed with water before being introduced to the pipeline; at the far end, the material must be dried. One example is a 525 km slurry pipeline which is planned to transport iron ore from the Minas-Rio mine (producing 26.5 million tonnes per year) to a port at Aηu in Brazil. An existing example is the 85 km Savage River Slurry pipeline in Tasmania, Australia, possibly the world's first when it was built in 1967. It includes a 366m bridge span at 167m above the Savage River.

Hydrogen Pipeline Transport

Figure: Hydrogen pipelines

Hydrogen pipeline transport is a transportation of hydrogen through a pipe as part of the hydrogen infrastructure.

Economics

Hydrogen pipeline transport is used to transport hydrogen from the point of production or delivery to the point of demand. Although the technology for hydrogen pipeline transport is proven, and the transport costs are similar to those of CNG, most hydrogen is produced in the place of demand, with an industrial production facility every 50 to 100 miles (80 to 160 km)

Hydrogen has problems with both hydrogen embrittlement and corrosion. Hydrogen has an active electron, and therefore behaves somewhat like a Halogen. For this reason, Hydrogen pipes have to resist corrosion, and are normally specified to be of expensive stainless steel. The problem is compounded because hydrogen can easily migrate into the crystal structure of most metals.

History

- 1938 - Rhine-Ruhr The first 240 km hydrogen pipes that are constructed of regular pipe steel, compressed hydrogen pressure 10-20 bar, diametre 25–30 cm. Still in operation.
- 1973 – 30 km pipeline in Isbergues, France.
- 1985 - Extension of the pipeline from Isbergues to Zeebrugge

- 1997 - Connection of the pipeline to Rotterdam
- 1997 - 2000: Development of two hydrogen networks, one near Corpus Christi, Texas, and one between Freeport and Texas City.
- 2009 - 150 mile extension of the pipeline from Plaquemine to Chalmette.

Infrastructure

- 2004 - USA - 900 miles (1450 km) of low pressure hydrogen pipelines
- 2004 - Europe - 930 miles (1500 km) of low pressure hydrogen pipelines.

Research

Fibre-Reinforced Polymer Pipelines (or FRP pipeline) and Reinforced thermoplastic pipes are researched.

For Water

Two millennia ago the ancient Romans made use of large aqueducts to transport water from higher elevations by building the aqueducts in graduated segments that allowed gravity to push the water along until it reached its destination. Hundreds of these were built throughout Europe and elsewhere, and along with flour mills were considered the lifeline of the Roman Empire. The ancient Chinese also made use of channels and pipe systems for public works. The famous Han Dynasty court eunuch Zhang Rang (d. 189 AD) once ordered the engineer Bi Lan to construct a series of square-pallet chain pumps outside the capital city of Luoyang. These chain pumps serviced the imperial palaces and living quarters of the capital city as the water lifted by the chain pumps was brought in by a stoneware pipe system.

Pipelines are useful for transporting water for drinking or irrigation over long distances when it needs to move over hills, or where canals or channels are poor choices due to considerations of evapouration, pollution, or environmental impact.

The 530 km (330 mi) Goldfields Water Supply Scheme in Western Australia using 750 mm (30 inch) pipe and completed in 1903 was the largest water supply scheme of its time.

Examples of significant water pipelines in South Australia are the Morgan-Whyalla (completed 1944) and Mannum-Adelaide (completed 1955) pipelines.

There are two Los Angeles, California aqueducts, the *First Los*

Angeles Aqueduct (completed 1913) and the *Second Los Angeles Aqueduct* (completed 1970) which also include extensive use of pipelines.

For Beverages

For Beer

Bars in the Veltins-Arena, a major football ground in Gelsenkirchen, Germany, are interconnected by a 5 km long beer pipeline. It is the favourite method for distributing beer in such large stadiums, because the bars have to overcome big differences between demands during various stages of a match; this allows them to be supplied by a central tank. In Randers city in Denmark, the so-called Thor beer pipeline still exists. Originally copper pipes were running directly from the brewery, and, when in 90s the brewery moved out the city, Thor beer replaced the centre of a star with a giant tank.

For Other Uses

The town of Hallstatt in Austria claims to contain "the oldest industrial pipeline in the world", dating back to 1595. It was constructed from 13,000 trunks to transport saline solution for 40 kilometres from Hallstatt to Ebensee.

Types by Transport Function

In general, pipelines can be classified in three categories depending on purpose:

Gathering Pipelines: Group of smaller interconnected pipelines forming complex networks with the purpose of bringing crude oil or natural gas from several nearby wells to a treatment plant or processing facility. In this group, pipelines are usually short- a couple of hundred metres- and with small diametres. Also sub-sea pipelines for collecting product from deep water production platforms are considered gathering systems.

Transportation Pipelines: Mainly long pipes with large diametres, moving products (oil, gas, refined products) between cities, countries and even continents. These transportation networks include several compressor stations in gas lines or pump stations for crude and multiproducts pipelines.

Distribution Pipelines: Composed of several interconnected pipelines with small diametres, used to take the products to the final consumer. Feeder lines to distribute gas to homes and businesses downstream. Pipelines at terminals for distributing products to tanks and storage facilities are included in this group.

Construction

When a pipeline is built, the construction project not only covers the civil work to lay the pipeline and build the pump/compressor stations, it also has to cover all the work related to the installation of the field devices that will support remote operation.

The pipeline is routed along what is known as a 'right of way'. Pipelines are generally built using the following stages:

1. Route (right of way) Selection
2. Surveying the route
3. Clearing the route
4. Trenching - Main Route and Crossings (roads, rail, other pipes, etc.)
5. Installing the pipe
6. Installing valves, intersections, etc.
7. Covering the pipe and trench.

Russia has Pipeline Troops as part of the Rear Services, who are trained to build and repair pipelines. Russia is the only country to have Pipeline Troops.

Operation

Field devices are instrumentation, data gathering units and communication systems. The field Instrumentation includes flow, pressure and temperature gauges/transmitters, and other devices to measure the relevant data required. These instruments are installed along the pipeline on some specific locations, such as injection or delivery stations, pump stations (liquid pipelines) or compressor stations (gas pipelines), and block valve stations.

The information measured by these field instruments is then gathered in local Remote Terminal Units (RTU) that transfer the field data to a central location in real time using communication systems, such as satellite channels, microwave links, or cellular phone connections.

Pipelines are controlled and operated remotely, from what is usually known as The Main Control Room. In this centre, all the data related to field measurement is consolidated in one central database. The data is received from multiple RTUs along the pipeline. It is common to find RTUs installed at every station along the pipeline.

The SCADA system at the Main Control Room receives all the field data and presents it to the pipeline operator through a set of screens or Human Machine Interface, showing the operational conditions of the pipeline.

The operator can monitor the hydraulic conditions of the line, as well as send operational commands (open/close valves, turn on/off compressors or pumps, change setpoints, etc.) through the SCADA system to the field.

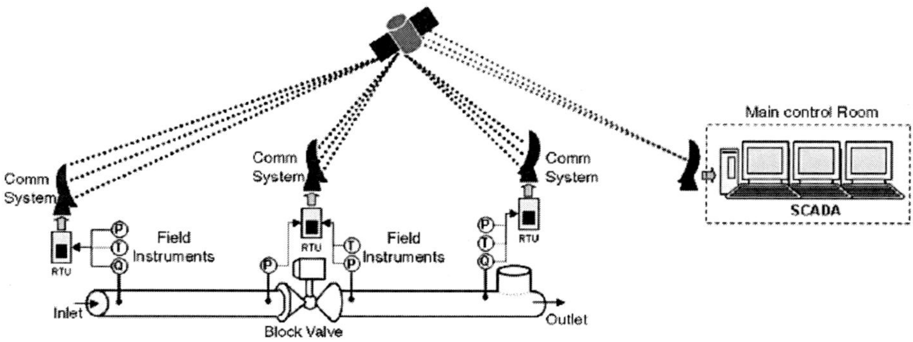

Figure: The SCADA System for pipelines.

To optimise and secure the operation of these assets, some pipeline companies are using what is called *Advanced Pipeline Applications*, which are software tools installed on top of the SCADA system, that provide extended functionality to perform leak detection, leak location, batch tracking (liquid lines), pig tracking, composition tracking, predictive modelling, look ahead modelling, operator training and more.

Technology

Components

Pipeline networks are composed of several pieces of equipment that operate together to move products from location to location. The main elements of a pipeline system are:

Initial Injection Station: Known also as *supply* or *inlet* station, is the beginning of the system, where the product is injected into the line. Storage facilities, pumps or compressors are usually located at these locations.

Compressor/Pump Stations: Pumps for liquid pipelines and Compressors for gas pipelines, are located along the line to move the product through the pipeline. The location of these stations is defined by the topography of the terrain, the type of product being transported, or operational conditions of the network.

Partial Delivery Station: Known also as *intermediate* stations, these facilities allow the pipeline operator to deliver part of the product being transported.

Block Valve Station: These are the first line of protection for pipelines. With these valves the operator can isolate any segment of the line for maintenance work or isolate a rupture or leak. Block valve stations are usually located every 20 to 30 miles (48 km), depending on the type of pipeline. Even though it is not a design rule, it is a very usual practice in liquid pipelines. The location of these stations depends exclusively on the nature of the product being transported, the trajectory of the pipeline and/or the operational conditions of the line.

Regulator Station: This is a special type of valve station, where the operator can release some of the pressure from the line. Regulators are usually located at the downhill side of a peak.

Final Delivery Station: Known also as outlet stations or terminals, this is where the product will be distributed to the consumer. It could be a tank terminal for liquid pipelines or a connection to a distribution network for gas pipelines.

Leak Detection Systems

Since oil and gas pipelines are an important asset of the economic development of almost any country, it has been required either by government regulations or internal policies to ensure the safety of the assets, and the population and environment where these pipelines run. Pipeline companies face government regulation, environmental constraints and social situations. Pipeline companies should comply with government regulations which may define minimum staff to run the operation, operator training requirements, up to specifics including pipeline facilities, technology and applications required to ensure operational safety. As an example, in the State of Washington, it is mandatory for pipeline operators to be able to detect and locate leaks of 8 percent of maximum flow within 15 minutes or less.

The social situation also affects the operation of pipelines. In third world countries, product theft is a problem for pipeline companies. It is common to find unauthorised extractions in the middle of the pipeline. In this case, the detection levels should be under 2 percent of maximum flow, with a high expectation for location accuracy.

Different types of technologies and strategies have been implemented, from physically walking the lines to satellite surveillance. The most common technology to protect these lines from occasional leaks is known as Computational Pipeline Monitoring Systems or CPM. CPM takes information from the field related to pressures, flows, and temperatures to estimate the hydraulic beahviour of the product being transported.

Once the estimation is done, the results are compared to other field references to detect the presence of an anomaly or unexpected situation, which may be related to a leak.

The American Petroleum Institute has published several articles related to the performance of CPM in liquids pipelines, the API Publications are:

- API 1130 – Computational pipeline monitoring for liquids pipelines
- API 1155 – Evaluation methodology for software based leak detection systems
- API 1149 – Pipeline variable uncertainties & their effects on leak detectability.

Implementation

As a rule pipelines for all uses are laid in most cases underground. However in some cases it is necessary to cross a valley or a river on a pipeline bridge. Pipelines for centralised heating systems are often laid on the ground or overhead. Pipelines for petroleum running through permafrost areas as Trans-Alaska-Pipeline are often run overhead in order to avoid melting the frozen ground by hot petroleum which would result in sinking the pipeline in the ground.

Regulation

In the US, onshore and offshore pipelines used to transport oil and gas are regulated by the Pipeline and Hazardous Materials Safety Administration (PHMSA). Certain offshore pipelines used to produce oil and gas are regulated by the Minerals Management Service (MMS). In Canada, pipelines are regulated by either the provincial regulators or, if they cross provincial boundaries or the Canada/US border, by the National Energy Board (NEB).

Government regulations in Canada and the United States require that buried fuel pipelines must be protected from corrosion. Often, the most economical method of corrosion control is by use of pipeline coating in conjunction with cathodic protection and technology to monitor the pipeline. Above ground, cathodic protection is not an option. The coating is the only external protection.

Pipelines and Geopolitics

Pipelines for major energy resources (petroleum and natural gas) are not merely an element of trade. They connect to issues of geopolitics and international security as well, and the construction, placement, and control of oil and gas pipelines. A notable example of pipeline politics occurred at the beginning of the year 2009, wherein a dispute between Russia and Ukraine ostensibly over pricing led to a major political crisis. Russian state-owned gas company Gazprom cut off natural gas supplies to Ukraine after talks between it and the Ukrainian government fell through. In addition to cutting off supplies to Ukraine, Russian gas flowing through Ukraine—which included nearly all supplies to Southeastern Europe and some supplies to Central and Western Europe—was cut off, creating a major crisis in several countries heavily dependent on Russian gas as fuel. Russia was accused of using the dispute as leverage in its attempt to keep other powers, and particularly the European Union, from interfering in its “near abroad”.

Dangers

Accidents: Pipelines conveying flammable or explosive material, such as natural gas or oil, pose special safety concerns.

- 1982 - One of the largest non-nuclear explosions in history occurred along the Trans-Siberian Pipeline in the former Soviet Union. It has been alleged that the explosion was the result of CIA sabotage of the Trans-Siberian Pipeline.
- June 4, 1989 - sparks from two passing trains detonated gas leaking from an LPG pipeline near Ufa, Russia. Up to 645 people were reported killed.
- October 17, 1998 - 1998 Jesse pipeline explosion at Jesse in the Niger Delta in Nigeria, a petroleum pipeline exploded killing about 1,200 villagers, some of whom were scavenging gasoline - the worst of several similar incidents in this country.
- June 10, 1999 - a pipeline rupture in a Bellingham, Washington park led to the release of 277,200 gallons of gasoline. The gasoline was ignited, causing an explosion that killed two children and one adult.
- August 19, 2000 - natural gas pipeline rupture and fire near Carlsbad, New Mexico; this explosion and fire killed 12 members of the same family. The cause was due to severe internal corrosion of the pipeline.

- July 30, 2004 - a major natural gas pipeline exploded in Ghislenghien, Belgium near Ath (thirty kilometres southwest of Brussels), killing at least 24 people and leaving 132 wounded, some critically. (Expatica)
- May 12, 2006 - an oil pipeline ruptured outside Lagos, Nigeria. Up to 200 people may have been killed.
- November 1, 2007 - a propane pipeline exploded near Carmichael, Mississippi, about 30 miles (48 km) south of Meridian, Mississippi. Two people were killed instantly and an additional four were injured. Several homes were destroyed and sixty families were displaced. The pipeline is owned by Enterprise Products Partners LP, and runs from Mont Belvieu, Texas, to Apex, North Carolina, according to an Enterprise spokesman.

As Targets

Pipelines can be the target of vandalism, sabotage, or even terrorist attacks. In war, pipelines are often the target of military attacks, as destruction of pipelines can seriously disrupt enemy logistics.

Hydraulically Activated Pipeline Pigging: Hydraulically activated pipeline pigging (HAPP) is a pigging technology applied for pipeline cleaning. The basic principle is that a pressure drop is created over a by-passable pig held back against a pipeline's fluid flow. The pipeline fluid passing through the pigs cleaning head is accelerated by this pressure drop forming strong cleaning jets.

These jets are directed onto the inner wall in front of the pig removing all kinds of deposits. Generally this technology transforms kinetic energy of the pipeline fluid into a locally available differential pressure which in this case is used to create cleaning jets but can also be used otherwise.

Introduction to Pipeline Pigging: Pipeline pigs are devices that are inserted into and travel throughout the length of a pipeline driven by the product flow. They were originally developed to remove deposits which could obstruct or retard flow through a pipeline. Today pigs are used during all phases in the life of a pipeline for many different reasons.

Pigs used today can be divided into three categories:

- Utility Pigs, are used to perform functions such as cleaning, separating, or dewatering.
- In Line Inspection (ILI) Tools, provide information on the

condition of the line, as well as the extent and location of any problems.

- Gel Pigs, are used in conjunction with conventional pigs to optimise pipeline dewatering, cleaning, and drying tasks.

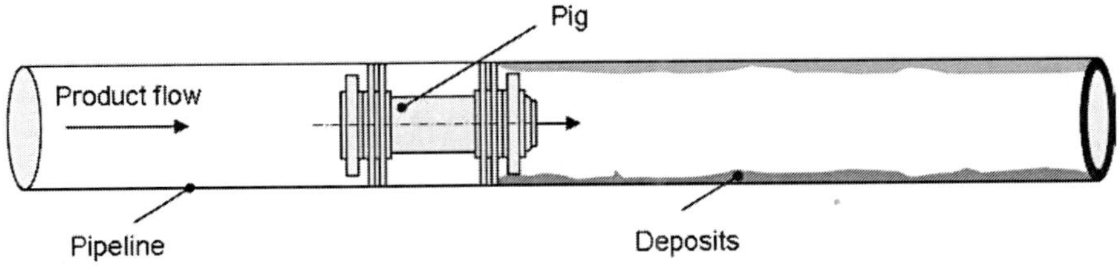

Figure: Cleaning pig in a pipeline

Generally for cleaning pigs, the cleaning force applied is the mechanical force between the pipe inner wall and the cleaning pig itself. This force is determined by the pig travel speed as well as by the hardness and shape of the cleaning edge: The faster the pig, the higher the cleaning impact on the deposits but at the same time only the surface of debris is scratched away. Therefore several, sometimes quite many, pig runs are required to clean a pipeline.

Hydraulic activated pigs apply high pressure liquid jets either supplied by high pressure hoses (depended) or made available by the kinetic energy locally available. Depended hydraulic activated pigs are limited in reach due to the hose which needs to be inserted into the pipeline and guides the cleaning head.

HAPP Principle

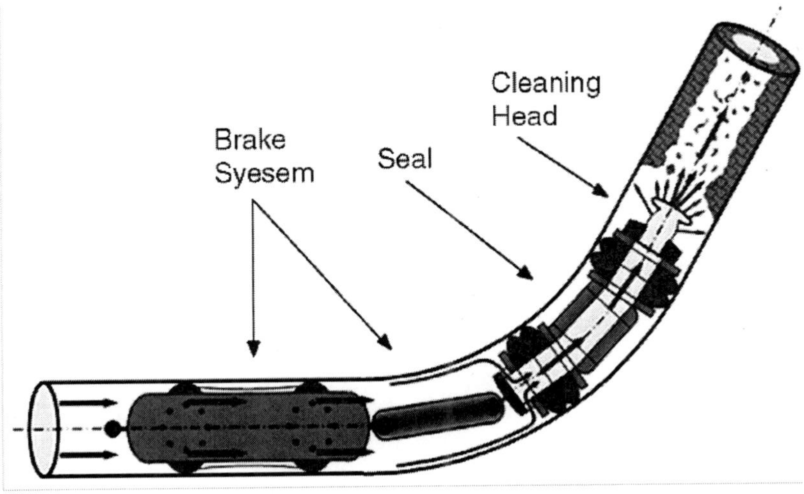

Figure: Components of a hydraulically activated power pig

A hydraulic activated pig consists of three units: a brake unit, a seal unit and the cleaning head. All units have openings that allow the entire fluid flow through the pipeline to bypass. The brake unit ensures that a hydraulically activated pig is held back against the fluid flow in the pipeline. The fluid pushes against the following seal unit, which channels it into the openings of the cleaning head.

The brake unit ensures that the travel speed of the pig is many times slower than the fluid velocity thus allowing it to entirely remove deposits from the pipe wall before it travels across the cleaned surface. The deposits removed are immediately flushed downstream the pipeline with the main jet of the cleaning head which is directed in the middle of the pipeline. With all deposits removed from the pipe wall and transported downstream by the fluid flow there remains no risk of the pig getting stuck in debris accumulated in front of it.

Airlift Pump

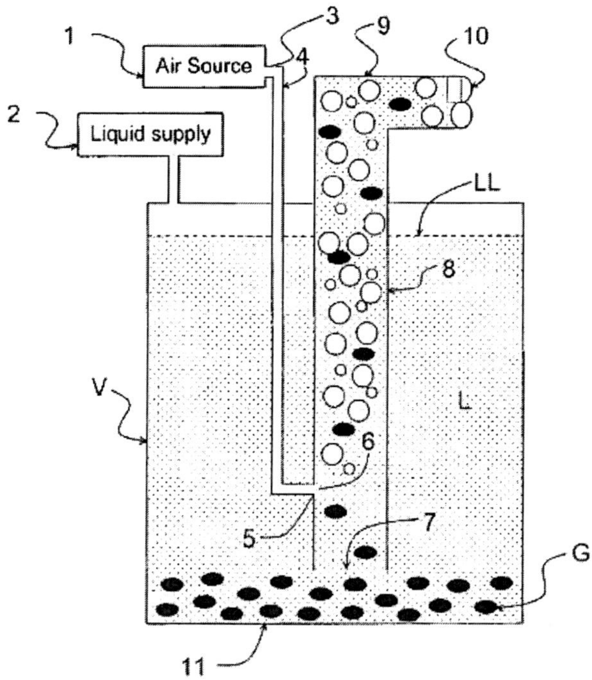

Figure: An airlift pump, powered by compressed air, raises fluid by entraining gas to reduce its density. 1. Air supply. 2. Liquid supply. 3. Air inlet port. 4. Air supply line. 5. Air port. 6. Air outlet. 7. Fluid intake. 8. Riser tube. 9. Air liquid mixture. 10. Pump outlet. L:Liquid, usually wastewater. LL:Liquid level. V:Vessel G:Gravel or solids.

An airlift pump is a simple pump which is powered by compressed air.

Principle

The only energy required is air. This air is usually compressed by a compressor or a blower. The air is injected in the lower part of a pipe that transports a liquid. It usually bubbles into another larger diametre pipe. By buoyancy the air, which has a lower density than the liquid, rises quickly. By fluid pressure, the liquid is taken in the ascendant air flow and moves in the same direction as the air. The calculation of the volume flow of the liquid is possible thanks to the physics of two-phase flow.

Airlift pump technology is superb due to its simple structure. However, it has the following weaknesses:

- Weak suction
- Unstable flow rate
- Frequent clogging
- Difficult flow control
- Low lift.

Utilisation

Airlift pumps are often used in deep dirty wells where sand would quickly abrade mechanical parts. (The compressor is on the surface and no mechanical parts are needed in the well). However airlift wells must be much deeper than the water table to allow for submergence. Air is generally pumped at least as deep under the water as the water is to be lifted. (If the water table is 50 ft below, the air should be pumped 100 feet deep). It is also sometimes used in part of the process on a wastewater treatment plant if a small head is required (typically around 1 foot head). They can also be used in ponds and aquaculture to aerate and mix the water.

Inventor

The first airlift pump is considered to be invented by the German engineer Carl Emanuel Löscher (de), who lived in the second part of the eighteenth century. He discovered the airlift pump in 1797.

Advantages and Disadvantages

The following paragraph exposes the advantages and disadvantages of the airlift pump compared to other pumping techniques.

Advantages

- The pump is very reliable. The very simple principle is a clear advantage. Only air with a higher pressure than the liquid is required.
- The liquid is not in contact with any mechanical elements. Therefore, neither the pump can be abraded (which is important for sandwater wells), neither the contents in the pipe can be abraded (which is important for archeological research in the sea).

Disadvantages

- cost: while in some specific case the operational cost can be interesting, most of the time, the quantity of air to compress is high compared to the liquid flow required.
- Conventional airlift pumps have a flow rate that is very limited. The pump is either on or off. It is very difficult to get a wide range of proportional flow control by varying the volume of compressed air. This is a dramatic disadvantage in some parts of a small wastewater treatment plant, such as the aerator.
- The suction is limited.
- This pumping system is suitable only if the head is relatively low. If you want to obtain a high head, you have to choose a conventional pumping system.
- Because of the principle, a lot of air remains in the liquid. In certain case, this can be problematic, as, for example, in a waste water treatment plant, before an anaerobic basin.

Design Improvements

A recent (2007) variant called the "geyser pump" can pump with greater suction and less air. It also pumps proportionally to the air flow, permitting use in processes that require varying controlled flows. It arranges to store up the air, and release it in large bubbles that seal to the lift pipe, raising slugs of fluid.

Back Pressure

Back pressure refers to the resistance to a moving fluid by obstructions or tight bends in the confinement vessel along which it is moving, such as piping or air vents, against its direction of flow.

Because it is really resistance, the term back pressure is misleading as the pressure remains and causes flow in the same direction, but the flow is reduced due to resistance. For example, an automotive exhaust muffler with a particularly high number of twists, bends, turns and right angles could be described as having particularly high back pressure.

Back pressure caused by the exhaust system (consisting of the exhaust manifold, catalytic converter, muffler and connecting pipes) of an automotive four-stroke engine has a negative effect on engine efficiency resulting in a decrease of power output that must be compensated by increasing fuel consumption; however, while the contemporary exhaust system may increase the level of strain on both the local economy and the global environment by requiring an increase in the purchase, consumption and venting of hydrocarbon fuel and its carbon dioxide byproduct, most automobile saturated societies see this as an acceptable and necessary trade-off. Although the exhaust system indirectly requires more fuel and that releases more CO_2 pollution into the upper atmosphere, the catalytic converter greatly reduces the heavier chemical byproducts produced by the engine that would otherwise remain much closer to the ground and disperse much more slowly causing a local pollution known as smog and with out mufflers, any area that is heavily trafficked by motor vehicles would become a highly dangerous environment for any people and perhaps animals as well who were not utilising professional hearing protection. However, the recent trend of making precision gas engines that run quieter and more efficient along with the current introduction of hybrids and fully electric cars is diminishing the importance of the exhaust system.

In a two-stroke engine however, a certain amount of exhaust back pressure is needed to prevent unburned fuel/air mixture from passing right through the cylinders into the exhaust.

Back Pressure in Information Technology

The term is also used analogously in the field of information technology to describe the build-up of data behind an I/O switch if the buffers are full and incapable of receiving any more data; the transmitting device halts the sending of data packets until the buffers have been emptied and are once more capable of storing information. It also refers to an algorithm for routing data according to congestion gradients.

Barlow's Formula

Barlow's formula relates the internal pressure that a pipe can withstand to its dimensions and the strength of its material.

$$P = \frac{2St}{D}$$

where

P = pressure (typically, pounds per square inch)

S = allowable stress (typically, pounds per square inch, yield or tensile depending upon application)

t = wall thickness (typically, inches)

D = outside diametre (typically, inches)

The formula is named after Peter Barlow, an English mathematician.

The design of a complex pressure containment system involves much more than the application of Barlow's formula. For almost all pressure vessels, the ASME code stipulates the requirements for design and testing.

Borda–Carnot Equation

In fluid dynamics the Borda–Carnot equation is an empirical description of the mechanical energy losses of the fluid due to a (sudden) flow expansion. It describes how the total head reduces due to the losses. This in contrast with Bernoulli's principle for dissipationless flow (without irreversible losses), where the total head is a constant along a streamline. The equation is named after Jean-Charles de Borda (1733–1799) and Lazare Carnot (1753–1823).

This equation is used both for open channel flow as well as in pipe flows. In parts of the flow where the irreversible energy losses are negligible, Bernoulli's principle can be used.

Formulation

The Borda–Carnot equation is:

$$\Delta E = \xi \frac{1}{2} \rho \left(v_1 - v_2 \right)^2,$$

where

- ΔE is the fluid's mechanical energy loss,
- ζ is an empirical loss coefficient, which is dimensionless and has a value between zero and one, $0 \leq \zeta \leq 1$,
- ρ is the fluid density,
- v_1 and v_2 are the mean flow velocities before and after the expansion.

In case of an abrupt and wide expansion the loss coefficient is equal to one. In other instances, the loss coefficient has to be determined by other means, most often from empirical formulae (based on data obtained by experiments).

The Borda–Carnot loss equation is only valid for decreasing velocity, $v_1 > v_2$, otherwise the loss ΔE is zero — without mechanical work by additional external forces there cannot be a gain in mechanical energy of the fluid.

The loss coefficient ζ can be influenced by streamlining. For example in case of a pipe expansion, the use of a gradual expanding diffuser can reduce the mechanical energy losses.

Relation to the Total Head and Bernoulli's Principle

The Borda–Carnot equation gives the decrease in the constant of the Bernoulli equation. For an incompressible flow the result is — for two locations labelled 1 and 2, with location 2 downstream to 1 — along a streamline:

$$p_1+\frac{1}{2}\rho v_1^2+\rho g z_1=p_2+\frac{1}{2}\rho v_2^2+\rho g z_2+\Delta E,$$

with

- p_1 and p_2 the pressure at location 1 and 2,
- z_1 and z_2 the vertical elevation — above some reference level of the fluid particle, and
- g the gravitational acceleration.

The first three terms, on either side of the equal sign are respectively the pressure, the kinetic energy density of the fluid and the potential energy density due to gravity. As can be seen, pressure acts effectively as a form of potential energy.

In case of high-pressure pipe flows, when gravitational effects can be neglected, ΔE is equal to the loss $\Delta(p+\frac{1}{2}\rho v^2)$:

$$\Delta E=\Delta\left(p+\frac{1}{2}\rho v^2\right).$$

For open channel flows, ΔE is related to the total head loss ΔH as:

$\Delta E=\rho g\Delta H$, with H the total head: $H=h+\dfrac{v^2}{2g}$,

where h is the hydraulic head — the free surface elevation above a reference datum: $h = z + p/(\rho g)$.

Examples

Sudden Expansion of a Pipe

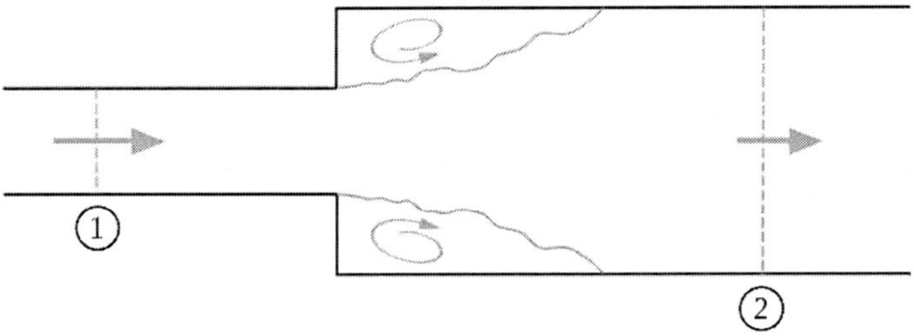

Figure: A sudden flow expansion.

The Borda–Carnot equation is applied to the flow through a sudden expansion of a horizontal pipe. The cross section 1, the mean flow velocity is equal to v_1, the pressure is p_1 and the cross-sectional area is A_1. The corresponding flow quantities at cross section 2 — after the expansion — are v_2, p_2 and A_2, respectively. The loss coefficient ζ for this sudden expansion is equal to one: $\zeta = 1.0$. Due to mass conservation, assuming a constant fluid density ρ, the volumetric flow rate through both cross sections 1 and 2 has to be equal: $A_1 v_1 = A_2 v_2$

so $v_2 = \frac{A_1}{A_2} v_1$.

Consequently — according to the Borda–Carnot equation — the mechanical energy loss in this sudden expansion is:

$$\Delta E = \frac{1}{2}\rho\left(1 - \frac{A_1}{A_2}\right)^2 v_1^2.$$

The corresponding loss of total head ΔH is:

$$\Delta H = \frac{\Delta E}{\rho g} = \frac{1}{2g}\left(1 - \frac{A_1}{A_2}\right)^2 v_1^2.$$

For this case with $\zeta = 1$, the total change in kinetic energy between the two cross sections is dissipated. As a result, the pressure change between both cross sections is (for this horizontal pipe without gravity effects):

$$\Delta p = p_1 - p_2 = -\rho\frac{A_1}{A_2}\left(1 - \frac{A_1}{A_2}\right)v_1^2,$$

and the change in hydraulic head $h = z + p/(\rho g)$:

$$\Delta h = h_1 - h_2 = -\frac{1}{g}\frac{A_1}{A_2}\left(1 - \frac{A_1}{A_2}\right)v_1^2.$$

The minus signs, in front of the right-hand sides, mean that the pressure (and hydraulic head) are larger after the pipe expansion. That this change in the pressures (and hydraulic heads), just before and after the pipe expansion, corresponds with an energy loss becomes clear when comparing with the results of Bernoulli's principle. According to this dissipationless principle, a reduction in flow speed is associated with a much larger increase in pressure than found in the present case with mechanical energy losses.

Sudden Contraction of a Pipe

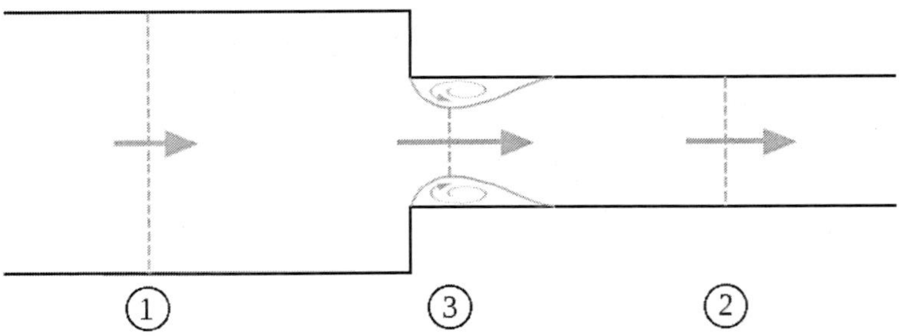

Figure: Flow through a sudden contraction of the pipe diametre, with flow separation bubbles near cross section 3.

In case of a sudden reduction of pipe diametre, without streamlining, the flow is not be able to follow the sharp bend into the narrower pipe. As a result, there is flow separation, creating recirculating separation zones at the entrance of the narrower pipe. The main flow is contracted between the separated flow areas, and later on expands again to cover the full pipe area.

There is not much head loss between cross section 1, before the contraction, and cross section 3, the vena contracta at which the main flow is contracted most. But there are substantial losses in the flow expansion from cross section 3 to 2. These head losses can be expressed by using the Borda–Carnot equation, through the use of the coefficient of contraction μ:

$$\mu = \frac{A_3}{A_2},$$

with A_3 the cross-sectional area at the location of strongest main flow contraction 3, and A_2 the cross-sectional area of the narrower part of the pipe. Since $A_3 \le A_2$, the coefficient of contraction is less than one: $\mu \leq 1$. Again there is conservation of mass, so the volume fluxes in the three cross sections are a constant (for constant fluid density ρ):

$$A_1 v_1 = A_2 v_2 = A_3 v_3,$$

with v_1, v_2 and v_3 the mean flow velocity in the associated cross sections. Then, according to the Borda–Carnot equation (with loss coefficient ζ=1), the energy loss ΔE per unit of fluid volume and due to the pipe contraction is:

$$\Delta E = \frac{1}{2}\rho\left(v_3 - v_2\right)^2 = \frac{1}{2}\rho\left(\frac{1}{\mu} - 1\right)^2 v_2^2 = \frac{1}{2}\rho\left(\frac{1}{\mu} - 1\right)^2 \left(\frac{A_1}{A_2}\right)^2 v_1^2.$$

The corresponding loss of total head ΔH can be computed as $\Delta H = \Delta E/(\rho g)$. According to measurements by Weisbach, the contraction coefficient for a sharp-edged contraction is approximately:

$$\mu = 0.63 + 0.37\left(\frac{A_2}{A_1}\right)^3.$$

Prony Equation

The Prony equation (named after Gaspard de Prony) is a historically important equation in hydraulics, used to calculate the head loss due to friction within a given run of pipe. It is an empirical equation developed by Frenchman Gaspard de Prony in the 19th century:

$$h_f = \frac{L}{D}(aV + bV^2)$$

where h_f is the head loss due to friction, calculated from: the ratio of the length to diametre of the pipe L/D, the velocity of the flow V, and two empirical factors a and b to account for friction.

This equation has been supplanted in modern hydraulics by the Darcy–Weisbach equation, which used it as a starting point.

British Standard Pipe Thread

The British Standard Pipe thread (BSP thread) is a family of standard screw thread types that has been adopted internationally for interconnecting and sealing pipe ends by mating an external (male) with an internal (female) thread.

Types

Two types of threads are distinguished:

- Parallel ('straight') threads, *British Standard Pipe Parallel thread* (*BSPP*; originally also known as *British Standard Pipe Fitting thread/BSPF* and *British Standard Pipe Mechanical thread/BSPM*), which have a constant diametre; denoted by the letter *G*.
- Taper threads, *British Standard Pipe Taper thread* (*BSPT*), whose diametre increases or decreases along the length of the thread; denoted by the letter *R*.

These can be combined into two types of joints:

- *Jointing threads:* These are pipe threads where pressure-tightness is made through the mating of two threads together. They always use a taper male thread, but can have either parallel or taper female threads. (In Europe, taper female pipe threads are not commonly used.)
- *Longscrew threads:* These are parallel pipe threads used where a pressure-tight joint is achieved by the compression of a soft material (such as an o-ring seal or a washer) between the end face of the male thread and a socket or nipple face, with the tightening of a backnut.

Thread Form

- the thread form follows the British Standard Whitworth standard
- symmetrical V-thread in which the angle between the flanks is 55° (measured in an axial plane)
- one-sixth of this sharp V is truncated at the top and the bottom
- the threads are rounded equally at crests and roots by circular arcs ending tangentially with the flanks
- the theoretical depth of the thread is therefore 0.64 times the nominal pitch

Pipe Thread Designations

These standard pipe threads are formally referred to by the following sequence of blocks:

- the words "Pipe thread",
- the document number of the standard (e.g., "ISO 7" or "EN 10226")
- the symbol for the pipe thread type:
  - G = external + internal parallel (ISO 228)

- R = external taper (ISO 7)
- Rp = internal parallel (ISO 7/1)
- Rc = internal taper (ISO 7)
- Rs = external parallel

- the thread size.

Threads are normally right-hand. For left-hand threads, the letters "LH" are appended.

Example: Pipe thread EN 10226 Rp 2½

The terminology for the use of *G* and *R* originated from Germany (*G* for gas, as it was originally designed for use on gas pipes; *R* for rohr (meaning pipe).)

Calibrated Orifice

A calibrated orifice is a restriction that is deliberately placed into a system of pipes to set the flow rate through the system. The orifice may be designed to produce proportional flow (as in the jet in a carburettor), or choked flow (as in a filtering bypass in a closed industrial cooling system, which might be designed to pass a particular flow rate through a filter assembly to maintain cleanliness of a closed-loop fluid system).

Many pressure gauges also use an orifice (also called a restrictor) to limit the flow into a gauge. Since the pressure is even throughout the system, allowing only a small portion of the flow into the actual gauge allows it to be in parallel with the pressure circuit and still measure accurately. It also prevents or minimises damage to the gauge during pressure surges at start-up, or due to any spikes in the system pressure.

Cast iron Pipe

Cast iron pipe is a historical pipe which found widespread use as a pressure pipe for transmission of water, gas and sewage, and as a water drainage pipe, during the 19th and 20th centuries. It comprises predominantly a gray cast iron tube and was frequently used uncoated, although later developments did result in various coatings and linings to reduce corrosion and improve hydraulics. Cast iron pipe was gradually superseded by ductile iron pipe, which is a direct development, with most existing manufacturing plants transitioning to the new material during the 1970s and 1980s. There is currently almost no new manufacture of cast iron pipe.

History

Early Use: The first use of cast iron pipe is not recorded. Cast iron tubes were first manufactured in the 14th century in Europe for cannon and it is presumed that similar tubes would have found use as water pipes at the same time, however, it was not until 1455 in Siegerland that the first officially recorded cast iron water pipe was produced for use in Dillenburg Castle (since destroyed).

The oldest extant water pipes date from the 17th century and were installed to distribute water throughout the gardens of the Chateau de Versailles. These amount to some 35 km of pipe, typically 1m lengths with flanged joints. The extreme age of these pipes make them of considerable historical value. Following extensive refurbishment in 2008 by Saint-Gobain PAM, 80% remain original.

Manufacture

Vertically Cast: The first cast iron pipe was produced in horizontal moulds which resulted in an uneven distribution of metal around the pipe circumference. In 1845, the first pipe was cast vertically and by the turn of the century, all pipe was manufactured to this method.

Centrifugally Cast

Subsequent to its invention by Dimitri Sensaud deLavaud, a French-Brazilian, in 1918, much cast iron pipe manufacture shifted to the dramatically different technique of centrifugal casting. Modern ductile iron pipe production continues to use this method of casting.

Cement-mortar Lined Ductile Iron Pipe

Cement-mortar lined ductile iron pipe is a ductile iron pipe with cement lining on the inside surface, and is commonly used for water distribution.

Cement-mortar lined ductile iron pipe is governed by standards set forth by DIPRA (Ductile Iron Pipe Research Association), and was first used in 1922 in Charleston, South Carolina.

Ductile Iron is commonly used in place of cast iron pipe for fluid distribution systems, the idea of lining the formerly cast iron and currently ductile iron was put in to practice for the first time in Charleston, South Carolina in 1922. The purpose of installing a cement/mortar lining to the interior wall of the pipe is to reduce the process of tuberculation inside the pipe network. The cement/mortar lining provides an area of high pH near the pipe wall and provides a barrier between the water and the pipe, reducing its susceptibility to corrosion.

Chézy Formula

In fluid dynamics, the Chézy formula describes the mean flow velocity of steady, turbulent open channel flow:

$$v = C\sqrt{Ri},$$

where

- u is the mean velocity [m/s],
- C is the Chézy coefficient [m½/s],
- R is the hydraulic radius (~ water depth) [m], and
- i is the bottom slope [m/m].

The formula is named after Antoine de Chézy, the French hydraulics engineer who devised it in 1775.

Usage with Manning Coefficient

This formula can also be used with Manning's Roughness Coefficient, instead of Chézy's coefficient. Manning derived the following relation to C based upon experiments:

$$C = \frac{1}{n} R^{1/6}$$

where

- C is the Chézy coefficient [m½/s],
- R is the hydraulic radius (~ water depth) [m], and
- n is Manning's roughness coefficient.

This relation is empirical.

Manning Formula

The Manning formula, known also as the Gauckler–Manning formula, or Gauckler–Manning–Strickler formula in Europe, is an empirical formula for open channel flow, or free-surface flow driven by gravity. It was first presented by the French engineer Philippe Gauckler in 1867, and later re-developed by the Irish engineer Robert Manning in 1890.

The Gauckler–Manning formula states:

$$V = \frac{k}{n} R_h^{2/3} \cdot S^{1/2}$$

where:

V is the cross-sectional average velocity (L/T; ft/s, m/s)

k is a conversion factor of 1 ($L^{1/3}/T$, $m^{1/3}/s$ for SI, or 1.4859 $ft^{1/3}/s$ U.S.

customary units, if required. Note: $(1\ \text{m})^{1/3}/\text{s} = (3.2808399\ \text{ft})^{1/3}/\text{s} = 1.4859\ \text{ft}^{1/3}/\text{s}$)

n is the Gauckler–Manning coefficient, it is unitless

R_h is the hydraulic radius (L; ft, m)

S is the slope of the water surface or the linear hydraulic head loss (L/L) ($S = h_f/L$)

The discharge formula, $Q = A\ V$, can be used to *manipulate* Gauckler–Manning's equation by substitution for V. Solving for Q then allows an estimate of the volumetric flow rate (discharge) without knowing the limiting or actual flow velocity.

The Gauckler–Manning formula is used to estimate flow in open channel situations where it is not practical to construct a weir or flume to measure flow with greater accuracy. The friction coefficients across weirs and orifices are less subjective than n along a natural (earthen, stone or vegetated) channel reach. Cross sectional area, as well as n', will likely vary along a natural channel. Accordingly, more error is expected in predicting flow by assuming a Manning's n, than by measuring flow across a constructed weirs, flumes or orifices. The formula can be obtained by use of dimensional analysis. Recently this formula was derived theoretically using the phenomenological theory of turbulence.

Hydraulic Radius

The hydraulic radius is a measure of a channel flow efficiency. Flow speed along the channel depends on its cross-sectional shape (among other factors), and the hydraulic radius is a characterisation of the channel that intends to capture such efficiency. Based on the 'constant shear stress at the boundary' assumption, hydraulic radius is defined as the ratio of the channel's cross-sectional area of the flow to its wetted perimetre (the portion of the cross-section's perimetre that is "wet"):

$$R_h = \frac{A}{P}$$

where:

R_h is the hydraulic radius (L),

A is the cross sectional area of flow (L^2),

P is wetted perimetre (L).

The greater the hydraulic radius, the greater the efficiency of the channel and the less likely the river is to flood. For channels of a given width, the hydraulic radius is greater for the deeper channels.

The hydraulic radius is *not* half the hydraulic diametre as the name may suggest. It is a function of the shape of the pipe, channel, or river in which the water is flowing. In wide rectangular channels, the hydraulic radius is approximated by the flow depth. The measure of a channel's efficiency (its ability to move water and sediment) is used by water engineers to assess the channel's capacity.

Gauckler–Manning Coefficient

The Gauckler–Manning coefficient, often denoted as n, is an empirically derived coefficient, which is dependent on many factors, including surface roughness and sinuosity. When field inspection is not possible, the best method to determine n is to use photographs of river channels where n has been determined using Gauckler–Manning's formula. In natural streams, n values vary greatly along its reach, and will even vary in a given reach of channel with different stages of flow. Most research shows that n will decrease with stage, at least up to bank-full. Overbank n values for a given reach will vary greatly depending on the time of year and the velocity of flow. Summer vegetation will typically have a significantly higher n value due to leaves and seasonal vegetation. Research has shown, however, that n values are lower for individual shrubs with leaves than for the shrubs without leaves. This is due to the ability of the plant's leaves to streamline and flex as the flow passes them thus lowering the resistance to flow. High velocity flows will cause some vegetation (such as grasses and forbs) to lay flat, where a lower velocity of flow through the same vegetation will not.

In open channels, the Darcy–Weisbach equation is valid using the hydraulic diametre as equivalent pipe diametre. It is the only sound method to estimate the energy loss in man-made open channels. For various reasons (mainly historical reasons), empirical resistance coefficients (e.g. Chézy, Gauckler–Manning–Strickler) were and are still used. The Chézy coefficient was introduced in 1768 while the *Gauckler–Manning coefficient was first developed in 1865, well before* the classical pipe flow resistance experiments in the 1920–1930s. Historically both the Chézy and the Gauckler–Manning coefficients were expected to be constant and functions of the roughness only. But it is now well recognised that these coefficients are only constant for a range of flow rates. Most friction coefficients (except perhaps the Darcy–Weisbach friction factor) are estimated 100% empirically and they apply only to fully rough turbulent water flows under steady flow conditions.

One of the most important applications of the Manning equation is its use in sewer design. Sewers are often constructed as circular pipes. It has long been accepted that the value of n varies with the flow depth in partially filled circular pipes. A complete set of explicit equations that can be used to calculate the depth of flow and other unknown variables when applying the Manning equation to circular pipes is available. These equations account for the variation of n with the depth of flow in accordance with the curves presented by Camp.

Closet Flange

Figure: PVC closet flange with steel flange, before installation

In plumbing, a closet flange (also known as a toilet flange) is a pipe fitting (specifically, a type of flange) that both mounts a toilet to the floor and connects the toilet drain to a drain pipe. The name comes from the term "water closet", the traditional name for a toilet. A typical closet flange is composed of an ABS or PVC hub with a round steel mounting flange attached to the top. Other styles are made from copper, brass, stainless steel, or pure plastic.

In a typical installation, the closet flange is mounted on top of the floor with the hub fused around the drain pipe. A wax ring is used to seal the gap between the flange and the bottom of the toilet. The toilet is bolted to the flange, not to the floor.

Double-walled Pipe

A double-walled pipe is a secondary contained piping system. It is a pipe within a pipe, or encased in an outer covering, with an annulus (interstitial space) between the two diametres. The inner pipe is the primary or carrier pipe and the outer pipe is called the secondary or containment pipe. The great majority of double-walled piping applications involve wastewater, groundwater, and process safety.

U.S. Environmental Protection Agency (EPA) standards, published in 40 CFR Parts 280 and 281, mandate double-walled piping for many below-ground transport systems in wastewater treatment plants and at sanitary or hazardous-waste landfills or remediation sites. These and other regulations affect below-grade transport of hazardous materials in fuel storage systems, tank farms, drainage or runoff from process plants, and some food-processing and related applications. Double-walled containment systems are also used for certain classes of toxic or corrosive chemicals, mostly gases, used in semiconductor fabrication. Double-walled pipes may also be used where a waste water pipe traverses a drinking water catchment area.

Double-walled pipe system may use plastic pressure pipe systems materials of high-density polyethylene (HDPE), polyvinyl chloride (PVC), chlorinated polyvinyl chloride (CPVC), polypropylene (PP), polyvinylidene fluoride (PVDF), and ethylene chlorotrifluoroethylene copolymer (ECTFE). Dual-wall stainless steel tubing is most commonly used in semiconductor plants for the containment of highly toxic gases.

A leak detection system in the containment pipe indicates if the carrier pipe is leaking. Double wall pipes are usually hydrostatically tested, or tested to hold and maintain a target gas pressure before the first use.

A different special application of a double-walled pipe is a jacketed pipe, which is used to make high viscosity liquids flow at elevated temperature through the carrier pipe. The outer pipe circulates hot fluids that heat up the interior carrier pipe and its contents in turn.

Drag Reducing Agent

A drag reducing agent, also called a flow improver, is a long chain polymer chemical that is used in crude oil, refined products or non-potable water pipelines. It is injected in small amounts (parts per million) and is used to reduce the frictional pressure drop along the pipeline's length.

The benefits of using a drag reducer are the following

1. Increase in pipeline throughput
2. Reduction of the waiting time for tanker loading/offloading
3. Maintaining the throughput during MOL (Main Oil Line) pump maintenance for de-rated lines

4. Bypassing MOL pump stations
5. Energy Savings.

The chemicals damp turbulent bursts of the oil near the pipeline wall, that way less disturbance is created during the oil flow. Minimising turbulence in the radial direction better preserves flow in the axial direction of the pipeline.

Drag reduction effectiveness for a given concentration is based on the turbulent characteristics of the pipeline. The maximum theoretical effect is the same as a pipe in laminar flow, where all of the turbulence is eliminated by the agent. Drag reduction effectiveness is measured as a percentage of the pipeline with no DRA present. For example, 75% drag reduction is representative of a pipeline that has one quarter of the frictional pressure loss at a given flow rate.

Since DRA is composed of long polymer strands, it is prone to degradation as it travels through the pipeline due to shearing of the strands. Large pressure changes through a control valve or pump result in a total loss of effectiveness. DRA may be reinjected after such equipment, but the total injection is usually limited by the product specifications or fluid limitations. DRA should never be used with any turbine fuels (such as jet fuel) because the polymer will accumulate on turbine blades and may damage the turbine.

Drag reducers were invented more than 30 years ago by Conoco Inc. (now ConocoPhillips). Its use has allowed pipeline systems to greatly increase in traditional capacity and extend the life of existing systems. The higher flow rates possible on long pipelines have also increased the potential for surge on older systems not previously designed for high velocities.

Drill Pipe

Drill pipe, is hollow, thick-walled, steel piping that is used on drilling rigs to facilitate the drilling of a wellbore and comes in a variety of sizes, strengths, and weights but are typically 30 to 33 feet in length. They are hollow to allow drilling fluid to be pumped through them, down the hole, and back up the annulus.

Because it is designed to support its own weight for combined lengths that often exceed 1 mile down into the Earth's crust, the case-hardened steel tubes are expensive, and owners spend considerable efforts to reuse them after finishing a well, replacing the drill stems with thinner walled tubular casing, tapping the natural resources of oil reservoirs. Used drill stem is often sent to a yard for inspection, sorted, and stored until new drill sites can be explored. Modified instruments similar to the spherometre are used at inspection sites to identify defects in the metallurgy, in order to prevent fracture of the drill stem during future wellboring.

Drill pipe is a portion of the overall drill string. The drill string consists of both drill pipe and the drill stem which is the tubular portion closest to the bit or downhole assembly. Drill pipe and drill stems can be differentiated in that the drill pipe is quite flexible and produced in longer segments whereas the drill stem is much more rigid and manufactured in shorter task specific segments.

Ductile Iron Pipe

Ductile iron pipe is a pipe commonly used for potable water distribution. The predominant wall material is ductile iron, a spheroidised graphite cast iron, although an internal cement mortar lining usually serves to inhibit corrosion from the fluid being distributed, and various types of external coating are used to inhibit corrosion from the environment. Ductile iron pipe is a direct development of earlier cast iron pipe which it has superseded. Ductile iron has proven to be a better pipe material, being stronger and more fracture resistant; however, like most ferrous materials, it is susceptible to corrosion and retains some brittle characteristics. Relatively recent developments such as polyethylene sleeving and/or zinc and polymer coatings have done much to mitigate corrosion concerns, and a typical life expectancy is in excess of 75 years.

Dimensions

Ductile iron pipe is sized according to a dimensionless term known as the Pipe Size or Nominal Diametre (known by its French abbreviation, DN). This is roughly equivalent to the pipe's internal diametre in inches or millimetres. However, it is the external diametre of the pipe that is kept constant between changes in wall thickness, in order to maintain compatibility in joints and fittings. Consequently the internal diametre varies, sometimes significantly, from its nominal size. Nominal pipe sizes vary from 3 inches up to 64 inches, in increments of at least 1 inch, in the USA. Pipe dimensions are standardised to the mutually incompatible AWWA C151 (U.S. Customary Units) in the USA, ISO 2531 / EN 545/598 (metric) in Europe, and AS/NZS 2280 (metric) in Australia and New Zealand. Although both metric, European and Australian are not compatible and pipes of identical nominal diametres have quite different dimensions.

North America

Pipe dimensions according to the American AWWA C-151

Pipe Size	Outside Diametre[in (mm)]
3	3.96 (100.584)
4	4.80 (121.92)
6	6.90 (175.26)
8	9.05 (229.87)
10	11.10 (281.94)
12	13.20 (335.28)
Pipe Size	**Outside Diametre[in (mm)]**
14	15.30 (388.62)
16	17.40 (441.96)
18	19.50 (495.3)
20	21.60 (548.64)
24	25.80 (655.32)
30	32.00 (812.8)

Joints

Individual lengths of ductile iron pipe are joined either by flanges, couplings, or some form of spigot and socket arrangement.

Flanges

Flanges are flat rings around the end of pipes which mate with an equivalent flange from another pipe, the two being held together by bolts usually passed through holes drilled through the flanges. A deformable gasket, usually elastomeric, placed between raised faces on the mating flanges provides the seal. Flanges are designed to a large number of specifications that differ because of dimensional variations in pipes sizes and pressure requirements, and because of independent standards development. In the U.S. flanges are either threaded or welded onto the pipe. In the European market flanges are usually welded on to the pipe. In the U.S. flanges are available in a standard 125 lb. bolt pattern as well as a 250 lb (and heavier) bolt pattern (steel bolt pattern). Both are usually rated at 250 psi (1,700 kPa). A flanged joint is rigid and can bear both tension and compression as well as a limited degree of shear and bending. It also can be dismantled after assembly. Due to the rigid nature of the joint and the risk of excessive bending moment being imposed, it is advised that flanged pipework is not buried.

Current flange standards used in the water industry are ANSI B16.1 in the USA, EN 1092 in Europe, and AS/NZS 4087 in Australia and New Zealand.

Spigot and Socket

Spigot and sockets involve a normal pipe end, the spigot, being inserted into the socket or bell of another pipe or fitting with a seal being made between the two within the socket. Normal spigot and socket joints do not allow direct metal to metal contact with all forces being transmitted through the elastomeric seal. They can consequently flex and allow some degree of rotation, allowing pipes to shift and relieve stresses imposed by soil movement. The corollary is that unrestrained spigot and socket joints transmit essentially no compression or tension along the axis of the pipe and little shear. Any bends, tees or valves therefore require either a restrained joint or, more commonly, thrust blocks, which transmit the forces as compression into the surrounding soil. A large number of different socket and seals exist. The most modern is the 'push-joint' or 'slip-joint', whereby the socket and rubber seal is designed to allow the pipe spigot to be, after lubrication, simply pushed into the socket. Push joints remain proprietary designs. Also available are locking gasket systems. These locking gasket systems allow the pipe to be pushed together but do not allow the joint to come apart without using a special tool or torch on the gasket. The earliest spigot and socket cast iron pipes were jointed by filling the socket with a mixture of water, sand, iron filings and sal-ammoniac (ammonium chloride.) A gaskin ring was pushed into the socket round the spigot to contain the mixture which was pounded into the socket with a caulking tool and then pointed off. This took several weeks to set and produced a completely rigid joint. Such pipe systems are often to be seen in nineteenth century churches in the heating system.

Manufacture

Ductile iron pipe is produced by a technique known as centrifugal casting, originally developed for cast iron pipe in 1918. The molten ductile iron is poured into a rapidly spinning water-cooled mould. Centrifugal force results in an even spread of iron around the circumference.

Internal Linings

Ductile iron pipe is somewhat resistant to internal corrosion in potable water and less aggressive forms of sewage. However, even where pipe material loss and consequently pipe wall reduction is slow, the deposition of corrosion products on the internal pipe wall can dramatically reduce the effective internal diametre and effectively choke flow, increasing pumping costs and lowering system pressure, long before the pipe itself is at risk of failure. A variety of linings are available to reduce or eliminate corrosion, including cement mortar, polyurethane and polyethylene. Of these, cement mortar lining is by far the most common.

Polyurethane (PUR)

Polyurethane protects piping made of ductile cast iron against corrosion and ensures meeting hygienic standards for drinking water at the same time. Polyurethane is used for both the inside lining and the outside coating. Because of polyurethane's elasticity, the coating remains intact even if the pipe is deformed. The PUR coating was developed in 1972. In comparison with other coatings, the internal polyurethane lining exhibits a high resistance to various different media such as drinking water, wastewater, de-mineralised water, industrial water and gas, as well as to aggressive solutions such as sulphuric acid. The PUR outside coating is suitable for all kinds of soil.

Specific Properties of Polyurethane

Composition Polyurethane is composed of a two-component resin. Its three-dimensionally linked molecular structure gives it its mechanical stability. Polyurethane is a thermosetting plastic with no solvents. It meets EN 545 and ISO 2531 standards.

Wall thickness Polyurethane PUR inside lining

DN 80-150 = 1.3 mm

DN 200-700 = 1.5 mm

Polyurethane PUR outside coating (ecopur®)

DN 80-700 = 0.9 mm

Colour Inside: green

Outside: black

Density 1.4 – 1.5 kg/dm3

Continuity continuous coating, no cracks

Bonding strength > 14 Mpa

(EMPA Recommends: 2.5 Mpa on a saturated sample)

Bonding strength is tested at regular intervals by our laboratories.

Dielectric resistance > 108 Ωm2

Temperature Water: up to 40°C (constant); up to 80°C (short term)

Air: 120°C

Impact resistance 40 Nm at 20°C

Effect of salt spray no effect after 1000 hours

Expansion > 10%

Friction coefficient k > 0.01 mm

Resistance to chemicals - Acidic or basic with pH values between 1 and 14

* Anorganic solvents
* Sulphuric acid (wasterwater)
* industrial wastewater

Thermal expansion 20 x 10-6 1/K

Deposits none : Use of chlorine The chlorine concentration in drinking water and the selective amount used.

when disinfecting have no influence on the quality of the polyurethane.

Quality Assurance Contract with EMPA in Dübendorf, Switzerland

Certification for PUR - Swiss Association for Gas and Water (SVGW)

* Swiss Federal Office of Health (BAG)
* Water Byelaws Advisory Service
* Singapore Institute of Standards and Industrial Research
* *for:* Bulgaria, Spain, Italy, Lithuania, Poland, Czech Republic, Rumania etc.

Cement Mortar

The predominant form of lining for water applications is cement mortar centrifugally applied during manufacturing. The cement mortar comprises a mixture of cement and sand to a ratio of between 1:2 and 1:3.5. For potable water, portland cement is used; for sewage it is common to use sulphate resisting or high alumina cement.

Cement mortar linings have been found to dramatically reduce internal corrosion. A DIPRA survey has demonstrated that the Hazen-Williams factor of cement lining remains between 130 and 151 with only slight reduction with age.

External Coatings

Unprotected ductile iron, similarly to cast iron, is intrinsically resistant to corrosion in most, although not all, soils. Nonetheless, because of frequent lack of information on soil aggressiveness and to extend the installed life of buried pipe, ductile iron pipe is commonly protected by one or more external coatings. In the U.S. and Australia, loose polyethylene sleeving is preferred. In Europe, standards recommend a more sophisticated system of directly bonded zinc coatings overlaid by a finishing layer be used in conjunction with polyethylene sleeving.

Polyethylene

Polyethylene sleeving was first developed by CIPRA (since 1979, DIPRA) in the U.S. in 1951 for use in highly corrosive soil in Birmingham, Alabama. It was employed more widely in the U.S. in the late 1950s and first employed in the UK in 1965 and Australia in the mid-1960s.

Polyethylene sleeving comprises a loose sleeve of polyethylene sheet that completely wraps the pipe, including the bells of any joints. Sleeving inhibits corrosion by a number of mechanisms. It physically separates the pipe from soil particles, preventing direct galvanic corrosion. By providing an impermeable barrier to ground water, the sleeve also inhibits the diffusion of oxygen to the ductile iron surface and limits the availability of electrolytes that would accelerate corrosion. It provides a homogeneous environment along the pipe surface so that corrosion occurs evenly over the pipe. The sleeve also restricts the availability of nutrients which could support sulfate-reducing bacteria, inhibiting microbially induced corrosion. Sleeving is not designed to be completely water-tight but rather to greatly restrict the movement of water to and from the pipe surface. Water present beneath the sleeve and in contact with the pipe surface is rapidly deoxygenated and depleted of nutrients and forms a stable environment in which limited further corrosion occurs. An improperly installed sleeve that continues to allow the free flow of ground water is not effective in inhibiting corrosion.

Polyethylene sleeves are available in a number of materials. The most common contemporary compositions are linear low-density polyethylene film which requires an 8 mil or 200 μm thickness and high-density cross-laminated polyethylene film which requires only a 4 mil or 100 μm thickness. The latter may or may not be reinforced with a scrim layer.

Polyethylene sleeving does have limitations. In European practice, its use in the absence of additional zinc and epoxy protective coatings is discouraged where natural soil resistivity is below 750 ohm/cm, where resistivity is below 1500 ohm/cm and where the pipe is installed at or below the water table, where there are additional artificial soil contaminants or where there are stray currents. Because of the vulnerability of polyethylene to UV degradation, sleeving, or sleeved pipe should not be stored in sunlight, although carbon pigments included in the sleeving can provide some limited protection.

Polyethylene sleeving is standardised according to ISO 8180 internationally, AWWA C105 in the U.S., BS 6076 in the UK and AS 3680 and AS 3681 in Australia.

Zinc

In Europe, ductile iron pipe is typically manufactured with a zinc coating overlaid by either a bituminous, polymer, or epoxy finishing layer. EN 545/598 mandates a minimum zinc content of 200 g/m^2 (with local minima of 110 g/m^2 at 99.99% purity) and a minimum average finishing layer thickness of 70 μm (with local minima of 50 μm).

No current AWWA standards are available for bonded coatings (zinc, coal tar epoxy, tape-wrap systems as seen on steel pipe) for ductile iron pipe, DIPRA does not endorse bonded coatings, and AWWA M41 generally views them unfavourably, recommending they be used only in conjunction with cathodic protection.

Bituminous Coatings

Zinc coatings are generally not employed in the U.S. and Australia. In order to protect ductile iron pipe prior to installation, pipe is instead supplied with a temporary 1 mil or 25 μm thick bituminous coating. This coating is not intended to provide protection once the pipe is installed.

Industry Associations

In the United States ductile iron pipe is often promoted to municipalities and consulting engineers by the Ductile Iron Pipe Research Association. Their focus is to promote the benefits of using ductile iron pipe on utility projects (water & sewer) over alternate products such as PVC, PCP, and HDPE.

Environmental

Ductile iron pipe in the developed world is normally manufactured exclusively from scrap steel. Ductile iron pipe can be recycled. In the

U.S. with the growing environmental movement ductile iron pipe is in a natural position to regain market share lost to its largest competitor, the PVC industry, over the past 40 years.

Electrofusion

Electrofusion is a method of joining MDPE, HDPE and other plastic pipes with special fittings that have built-in electric heating elements which are used to weld the joint together.

The pipes to be joined are cleaned, inserted into the electrofusion fitting (with a temporary clamp if required) and a voltage (typically 40V) is applied for a fixed time depending on the fitting in use, the built in heater coils then melt the inside of the fitting and the outside of the pipe wall which weld together producing a very strong homogeneous joint. The assembly is then left to cool for a specified time. The joints produced tend to be more reliable than threaded fittings sealed with O-rings.

External Water Spray System

An External Water Spray System (EWSS) is a domestic external fire sprinkler system designed to protect homes from bushfires and wildfires. While external spray systems have long been used in fire protection for buildings and facilities, EWSS refers to domestic bushfire/ wildfire systems.

Usage

The types of systems vary greatly from a single impact sprinkler placed on a roof, systems installed during construction with sprays on all windows and doors, and small sprays damping gutters. Some are installed using copper piping and sprays while others use common PVC piping.

Usage depends on type of risk and belief of effectiveness. Copper piping is used to withstand high temperatures that may be experienced during a fire front or for higher reliability for in ceiling installations. External PVC piping is used where failure with exposure to high radiant heat is acceptable, as at the time the system is considered have "done its job". This is a matter of personal judgment.

Issues

There is a lack of scientific research regarding EWSS. Other issues that may affect the effectiveness of an EWSS:

- Sprinkler performance in high wind conditions typical in a bushfire

- Home design
  - o High vulnerability areas. e.g. Decking
  - o Garden beds near walls
  - o Window sills
  - o Large windows (heat radiation)
  - o Internal furnishings (e.g. curtains).
- Home location
  - o Slope
  - o Surrounding vegetation
- If defended by occupiers
- Preparation (if prepared for defence)
- If intended for remote (unattended) defence
- Available water supply
- Ember attack
- Fire front (radiant heat)
- Adjacent building or vegetation fire.

Types

- Roof impact sprinkler
- Gutter sprays
- Window spray / deluge.

Commercial sprays designed for EWSS are available, but most systems use commonly available irrigation sprinklers/sprays.

Research

External spray systems for buildings are well documented for protection from fires in adjacent buildings. However, up there is little published scientific research on scientific information pertaining to the effectiveness of EWSS under varying wildfire conditions.

Fanning Friction Factor

The Fanning friction factor, named after John Thomas Fanning (1837–1911), is a dimensionless number used in fluid flow calculations. It is related to the shear stress at the wall as:

$$\tau = \frac{f \rho v^2}{2}$$

where:

- T is the shear stress at the wall
- f is the Fanning friction factor of the pipe
- u is the fluid velocity in the pipe
- p is the density of the fluid

The wall shear stress can, in turn, be related to the pressure loss by multiplying the wall shear stress by the wall area ($2\pi RL$ for a pipe) and dividing by the cross-sectional flow area (πR^2 for a pipe).

The friction head can be related to the pressure loss due to friction by dividing the pressure loss by the product of the acceleration due to gravity and the density of the fluid. Accordingly, the relationship between the friction head and the Fanning friction factor is:

$$h_f = \frac{2fv^2L}{gD}$$

where:

- h_f is the friction loss (in head) of the pipe.
- f is the Fanning friction factor of the pipe.
- u is the fluid velocity in the pipe.
- L is the length of pipe.
- g is the local acceleration of gravity.
- D is the pipe diametre.

Fanning Friction Factor Formulæ

This friction factor is one-fourth of the Darcy friction factor, so attention must be paid to note which one of these is meant in the "friction factor" chart or equation consulted. Of the two, the Fanning friction factor is the more commonly used by Chemical Engineers and those following the British convention.

The formulae below may be used to obtain the Fanning friction factor for common applications. The friction factor for laminar flow in round tubes is often taken to be:

$$f = \frac{16}{Re}$$

where Re is the Reynolds number of the flow.

For a square channel the value used is:

$$f = \frac{14.227}{Re}$$

The Darcy friction factor can also be expressed as

$$f = \frac{8\tau_w}{\rho V_{avg}^2}$$

where:

- τ_w is the shear stress at the wall
- p is the density of the fluid
- V_{avg}^2 is the average fluid velocity

For the turbulent flow regime, the relationship between the Fanning friction factor and the Reynolds number is more complex and is governed by the Colebrook equation which is implicit in f:

$$\frac{1}{\sqrt{f}} = -4.0\log_{10}\left(\frac{\frac{\epsilon}{d}}{3.7} + \frac{1.256}{Re\sqrt{f}}\right), \text{turbulent flow}$$

Various explicit approximations of the related Darcy friction factor have been developed for turbulent flow.

Stuart W. Churchill developed a formula that covers the friction factor for both laminar and turbulent flow.

$$f = 8\left(\left(\frac{8}{Re}\right)^{12} + \left(A+B\right)^{-1.5}\right)^{\frac{1}{12}}$$

$$A = \left(-2.457\ln\left(\left(\frac{7}{Re}\right)^{0.9} + 0.27\frac{e}{D}\right)\right)^{16}$$

$$B = \left(\frac{37530}{Re}\right)^{16}$$

Fire Sprinkler System

A fire sprinkler system is an active fire protection measure, consisting of a water supply system, providing adequate pressure and flowrate to a water distribution piping system, onto which fire sprinklers are connected. Although historically only used in factories and large commercial buildings, home and small building systems are now available at a cost-effective price.

Figure: A glass bulb type sprinkler head will spray water into the room if sufficient heat reaches the bulb and causes it to shatter. Sprinkler heads operate individually. Note the red liquid in the glass bulb.

History

The world's first recognisable sprinkler system was installed in the Theatre Royal, Drury Lane in the United Kingdom in 1812 by its architect. The system was designed by Sir William Congreve, 2nd Baronet was covered by a patent No 3606 dated the same year. The apparatus consisted of a cylindrical airtight reservoir of 400 hogsheads (~95,000 litres) fed by a 10 inches (250 mm) water main which branched to all parts of the theatre. A series of smaller pipes feed from the distribution pipe were pierced with a series of $^1/_2$ inches (13 mm) holes which pour water in the event of a fire.

From 1852 to 1885, perforated pipe systems were used in textile mills throughout New England as a means of fire protection. However, they were not automatic systems; they did not turn on by themselves. Inventors first began experimenting with automatic sprinklers around 1860. The first automatic sprinkler system was patented by Philip W. Pratt of Abington, MA, in 1872.

Henry S. Parmalee of New Haven, Connecticut is considered the inventor of the first automatic sprinkler head. Parmalee improved upon the Pratt patent and created a better sprinkler system. In 1874, he installed his fire sprinkler system into the piano factory that he owned. Frederick Grinnell improved Parmalee's design and in 1881 patented the automatic sprinkler that bears his name. He continued to improve the device and in 1890 invented the glass disc sprinkler, essentially the same as that in use today.

Until the 1940s, sprinklers were installed almost exclusively for the protection of commercial buildings, whose owners were generally able to recoup their expenses with savings in insurance costs. Over the years, fire sprinklers have become mandatory safety equipment in some parts of North America, in certain occupancies, including, but not limited to newly constructed hospitals, schools, hotels and other public buildings, subject to the local building codes and enforcement. However, outside of the US and Canada, sprinklers are rarely mandated by building codes for normal hazard occupancies which do not have large numbers of occupants (e.g. factories, process lines, retail outlets, petrol stations etc.)

Usage

Sprinklers have been in use in the United States since 1874, and were used in factory applications where fires at the turn of the century were often catastrophic in terms of both human and property losses. In the US, sprinklers are today required in all new high rise and underground buildings generally 75 feet (23 m) above or below fire department access, where the ability of firefighters to provide adequate hose streams to fires is limited.

Sprinklers may be required to be installed by building codes, or may be recommended by insurance companies to reduce potential property losses or business interruption. Building codes in the United States for places of assembly, generally over 100 persons, and places with overnight sleeping accommodation such as hotels, nursing homes, dormitories, and hospitals usually require sprinklers either under local building codes, as a condition of receiving State and Federal funding or as a requirement to obtain certification (essential for institutions who wish to train medical staff). Since 2011, Pennsylvania and California require sprinkler systems in all new residential construction-the first US states to do so.

Federal Law (US)

While there is very little specific Federal legislation regarding building codes, which are generally left to local jurisdictions, the Federal government has used its funding and monetary clout to strongly encourage fire safety standards in construction. Since 2011, two states, California and Pennsylvania have required sprinklers in at least some new residential construction.

In 1990 the US Congress passed PL-101-391, better known as "The Hotel and Motel Safety Act (of 1990)". This law requires that any hotel, meeting hall, or similar institution that receives federal funds (i.e. for a government traveller's overnight stay, or a conference, etc.), must meet fire and other safety requirements. The most visible of these conditions are the implementation of sprinklers.

As more and more hotels, etc., upgraded their facilities to enable acceptance of government visitors, this type of construction became the de facto industry norm-even when not directly mandated by any local building codes.

If building codes do not explicitly mandate the use of fire sprinklers, the code often makes it highly advantageous to install them as an optional system. Most US building codes allow for less expensive construction materials, larger floor area limitations, longer egress paths, and fewer requirements for fire rated construction in structures protected by fire sprinklers. Consequently, the total building cost is often less by installing a sprinkler system and savings money in the other aspects of the project, as compared to building a non-sprinklered structure. In the UK, since the 1990s sprinklers have gained recognition within the Building Regulations (England and Wales) and Scottish Building Standards and under certain circumstances, the presence of sprinkler systems is deemed to provide a form of alternative compliance to some parts of the codes. For example, the presence of a sprinkler system will usually permit doubling of compartment sizes and increases in travel distances (to fire exits) as well as allowing a reduction in the fire rating of internal compartment walls.

Europe

Renewed interest in and support for sprinkler systems in the UK, largely as a result of effective lobbying by the National Fire Sprinkler Network, the European Fire Sprinkler Network and the British Automatic Fire Sprinkler Association, has resulted in sprinkler systems being more widely installed. In schools, for example, the government has issued recommendations through Building Bulletin 100 that most new schools should be constructed with sprinkler protection. In 2011 Wales became the first country in the world where sprinklers are compulsory in all new homes.

The law will apply to newly built houses and blocks of flats, as well as care homes and university halls of residence. In Scotland, all new schools are sprinklered, as are new care homes, sheltered housing and high rise flats. In Norway as of July 2010, all new housing of more than two storeys, all new hotels, care homes and hospitals must be sprinklered. Other Nordic countries require or soon will require sprinklers in new care homes, and in Finland as of 2010 a third of care homes were retrofitted with sprinklers. A fire in an illegal immigrant detention centre at Schiphol airport in The Netherlands on 27 October 2005 killed 11 detainees and led to the retrofitting of sprinklers in all similarly designed prisons in The Netherlands. A fire at Düsseldorf airport on 11 April 1996 which killed 17 people led to sprinklers being retrofitted in all major German airports. Most European countries also require sprinklers in shopping centres, in large warehouses and in high-rise buildings.

Operation

Each closed-head sprinkler is held closed by either a heat-sensitive glass bulb or a two-part metal link held together with fusible alloy. The glass bulb or link applies pressure to a pip cap which acts as a plug which prevents water from flowing until the ambient temperature around the sprinkler reaches the design activation temperature of the individual sprinkler head. In a standard wet-pipe sprinkler system, each sprinkler activates independently when the predetermined heat level is reached. Because of this, the number of sprinklers that operate is limited to only those near the fire (in reality, normally one or two will activate), thereby maximising the available water pressure over the point of fire origin. This also minimises the water damage to the building.

A sprinkler activation will do less damage than a fire department hose stream, which provide approximately 900 liters/min (250 US gallons/min). A typical sprinkler used for industrial manufacturing occupancies discharge about 75-150 litres/min (20-40 US gallons/min). However, a typical Early Suppression Fast Response (ESFR) sprinkler at a pressure of 50 psi (340 kPa) will discharge approximately 100 US gallons per minute (0.0063 m^3/s). In addition, a sprinkler will usually activate between one and four minutes, whereas the fire department typically takes at least five minutes to arrive at the fire site after receiving an alarm, and an additional ten minutes to set up equipment and apply hose streams to the fire. This additional time can result in a much larger fire, requiring much more water to extinguish.

Types

Wet Pipe Systems

By a wide margin, wet pipe sprinkler systems are installed more often than all other types of fire sprinkler systems. They also are the most reliable, because they are simple, with the only operating components being the automatic sprinklers and (commonly, but not always) the automatic alarm check valve. An automatic water supply provides water under pressure to the system piping.

Dry Pipe Systems

Dry pipe systems are installed in spaces in which the ambient temperature may be cold enough to freeze the water in a wet pipe system, rendering the system inoperable. Dry pipe systems are most often used in unheated buildings, in parking garages, in outside canopies attached to heated buildings (in which a wet pipe system would be provided), or in refrigerated coolers. Dry pipe systems are the second most common sprinkler system type. In regions using NFPA regulations, dry pipe systems cannot be installed unless the range of ambient temperatures reaches below 40F.

Operation-Water is not present in the piping until the system operates. The piping is filled with air below the water supply pressure. To prevent the larger water supply pressure from forcing water into the piping, the design of the dry pipe valve (a specialised type of check valve) results in a greater force on top of the check valve clapper by the use of a larger valve clapper area exposed to the piping air pressure, as compared to the higher water pressure but smaller clapper surface area.

When one or more of the automatic sprinklers is exposed, for a sufficient time, to a temperature at or above the temperature rating, it opens, allowing the air in the piping to vent from that sprinkler. Each sprinkler operates individually. As the air pressure in the piping drops, the pressure differential across the dry pipe valve changes, allowing water to enter the piping system. Water flow from sprinklers, needed to control the fire, is delayed until the air is vented from the sprinklers. In regions using NFPA 13 regulations, the time it takes water to reach the hydraulically remote sprinkler from the time that sprinkler is activated is limited to a maximum of 60 seconds. In industry practice, this is known as the "Maximum Time of Water Delivery." The maximum time of water delivery may be required to be reduced, depending on the hazard classification of the area protected by the sprinkler system.

Some view dry pipe sprinklers as advantageous for protection of collections and other water sensitive areas. This perceived benefit is due to a fear that wet system piping may leak, while dry pipe systems will not. However, the same potential for accidental water damage exists, as dry pipe systems will only provide a slight delay prior to water discharge while the air in the piping is released from the pipe.

Disadvantages of using dry pipe fire sprinkler systems include:

- Increased complexity-Dry pipe systems require additional

control equipment and air pressure supply components which increases system complexity. This puts a premium on proper maintenance, as this increase in system complexity results in an inherently less reliable overall system (i.e., more single failure points) as compared to a wet pipe system.

- Higher installation and maintenance costs-The added complexity impacts the overall dry-pipe installation cost, and increases maintenance expenditure primarily due to added service labour costs.
- Lower design flexibility-Regulatory requirements limit the maximum permitted size (i.e., 750 gallons) of individual dry-pipe systems, unless additional components and design efforts are provided to limit the time from sprinkler activation to water discharge to under one minute. These limitations may increase the number of individual sprinkler zones (i.e., served from a single riser) that must be provided in the building, and impact the ability of an owner to make system additions.
- Increased fire response time-Because the piping is empty at the time the sprinkler operates, there is an inherent time delay in delivering water to the sprinklers which have operated while the water travels from the riser to the sprinkler, partially filling the piping in the process. A maximum of 60 seconds is normally allowed by regulatory requirements from the time a single sprinkler opens until water is discharged onto the fire. This delay in fire suppression results in a larger fire prior to control, increasing property damage.
- Increased corrosion potential-Following operation or testing, dry-pipe sprinkler system piping is drained, but residual water collects in piping low spots, and moisture is also retained in the atmosphere within the piping. This moisture, coupled with the oxygen available in the compressed air in the piping, increases pipe internal wall corrosion rates, possibly eventually leading to leaks. The internal pipe wall corrosion rate in wet pipe systems (in which the piping is constantly full of water) is much lower, as the amount of oxygen available for the corrosion process is lower. Corrosion can be combated by using galvanised steel pipe which is less susceptible to corrosion, or by using nitrogen to pressurise the system, rather than air. These additional precautions will increase the cost of the system, but can help prevent system failure and premature need for system replacement in the future.

Deluge Systems

"Deluge" systems are systems in which all sprinklers connected to the water piping system are open, in that the heat sensing operating element is removed, or specifically designed as such. These systems are used for special hazards where rapid fire spread is a concern, as they provide a simultaneous application of water over the entire hazard. They are sometimes installed in personnel egress paths or building openings to slow travel of fire (e.g., openings in a fire-rated wall).

Water is not present in the piping until the system operates. Because the sprinkler orifices are open, the piping is at atmospheric pressure. To prevent the water supply pressure from forcing water into the piping, a *deluge valve* is used in the water supply connection, which is a mechanically latched valve. It is a non-resetting valve, and stays open once tripped.

Because the heat sensing elements present in the automatic sprinklers have been removed (resulting in open sprinklers), the deluge valve must be opened as signalled by a fire alarm system. The type of fire alarm initiating device is selected mainly based on the hazard (e.g., smoke detectors, heat detectors, or optical flame detectors). The initiation device signals the fire alarm panel, which in turn signals the deluge valve to open. Activation can also be manual, depending on the system goals. Manual activation is usually via an electric or pneumatic fire alarm pull station, which signals the fire alarm panel, which in turn signals the deluge valve to open.

Operation-Activation of a fire alarm initiating device, or a manual pull station, signals the fire alarm panel, which in turn signals the deluge valve to open, allowing water to enter the piping system. Water flows from all sprinklers simultaneously.

Pre-Action Systems

Pre-action sprinkler systems are specialised for use in locations where accidental activation is undesired, such as in museums with rare art works, manuscripts, or books; and Data Centres, for protection of computer equipment from accidental water discharge.

Pre-action systems are hybrids of wet, dry, and deluge systems, depending on the exact system goal. There are two main sub-types of pre-action systems: single interlock, and double interlock.

The operation of single interlock systems are similar to dry systems except that these systems require that a "preceding" fire detection event, typically the activation of a heat or smoke detector, takes place prior to the "action" of water introduction into the system's piping by opening the pre-action valve, which is a mechanically latched valve (i.e., similar to a deluge valve). In this way, the system is essentially converted from a dry system into a wet system. The intent is to reduce the undesirable time delay of water delivery to sprinklers that is inherent in dry systems. Prior to fire detection, if the sprinkler operates, or the piping system develops a leak, loss of air pressure in the piping will activate a trouble alarm. In this case, the pre-action valve will not open due to loss of supervisory pressure, and water will not enter the piping.

The operation of double interlock systems are similar to deluge systems except that automatic sprinklers are used. These systems require that both a "preceding" fire detection event, typically the activation of a heat or smoke detector, and an automatic sprinkler operation take place prior to the "action" of water introduction into the system's piping. Activation of either the fire detectors alone, or sprinklers alone, without the concurrent operation of the other, will not allow water to enter the piping. Because water does not enter the piping until a sprinkler operates, double interlock systems are considered as dry systems in terms of water delivery times, and similarly require a larger design area.

Foam Water Sprinkler Systems

A foam water fire sprinkler system is a special application system, discharging a mixture of water and low expansion foam concentrate, resulting in a foam spray from the sprinkler. These systems are usually used with special hazards occupancies associated with high challenge fires, such as flammable liquids, and airport hangars. Operation is as described above, depending on the system type into which the foam is injected.

Water Spray

"Water spray" systems are operationally identical to a deluge system, but the piping and discharge nozzle spray patterns are designed to protect a uniquely configured hazard, usually being three dimensional components or equipment (i.e., as opposed to a deluge system, which is designed to cover the horizontal floor area of a room). The nozzles used may not be listed fire sprinklers, and are usually selected for a specific spray pattern to conform to the three dimensional nature of the hazard (e.g., typical spray patterns being oval, fan, full circle, narrow jet). Examples of hazards protected by water spray systems are electrical transformers containing oil for cooling or turbo-generator bearings. Water spray systems can also be used externally on the surfaces of tanks containing flammable liquids or gases (such as hydrogen). Here the water spray is intended to cool the tank and its contents to prevent tank rupture/explosion (BLEVE) and fire spread.

Water Mist Systems

Water mist systems are used for special applications in which it is decided that creating a heat absorbent vapour is the primary primary objective. This type of system is typically used where water damage may be a concern, or where water supplies are limited. NFPA 750 defines water mist as a water spray with a droplet size of "less than 1000 microns at the minimum operation pressure of the discharge nozzle." The droplet size can be controlled by the adjusting discharge pressure through a nozzle of a fixed orifice size. By creating a mist, and equal volume of water will create a larger total surface area exposed to the fire. The larger total surface area better facilitates the transfer of heat, thus allowing more water droplets to turn to steam more quickly. Steam, which absorbs more heat than water per unit time, will then more effectively cool the room, thus reducing the temperature of the flame.

Operation-Water mist systems can operate with the same functionality as deluge, wet pipe, dry pipe, or pre-action systems. The difference is that a water mist system uses a compressed gas as an atomizing medium, which is pumped through the sprinkler pipe. Systems can be applied using local application method or total flooding method, similar to Clean Agent Fire Protection Systems.

Design

	Temperature	*Colour of Liquid inside Bulb*
°C	*°F*	*Orange*
57	135	Red
68	155	Yellow
79	174	Green
93	200	Blue
141	286	Mauve
182	360	Black
227260	440500	

New Zealand fire safety standards indicates the colour of the bulb and the respective operating temperature.

Sprinkler systems are intended to either *control* the fire or to *suppress* the fire. *Control mode* sprinklers are intended to control the heat release rate of the fire to prevent building structure collapse, and pre-wet the surrounding combustibles to prevent fire spread. The fire is not extinguished until the burning combustibles are exhausted or manual extinguishment is effected by firefighters. *Suppression mode* sprinklers (formerly known as *Early Suppression Fast Response* (ESFR) sprinklers) are intended to result in a severe sudden reduction of the heat release rate of the fire, followed quickly by complete extinguishment, prior to manual intervention.

Most sprinkler systems installed today are designed using an area and density approach. First the building use and building contents are analysed to determine the level of fire hazard. Usually buildings are classified as light hazard, ordinary hazard group 1, ordinary hazard group 2, extra hazard group 1, or extra hazard group 2. After determining the hazard classification, a design area and density can be determined by referencing tables in the National Fire Protection Association (NFPA) standards. The design area is a theoretical area of the building representing the worst case area where a fire could burn. The design density is a measurement of how much water per square foot of floor area should be applied to the design area. For example, in an office building classified as light hazard, a typical design area would be 1,500 square feet (140 m^2) and the design density would be 0.1 US gallons per minute (6.3×10^{-6} m^3/s) per 1 square foot (0.093 m^2) or a minimum of 150 US gallons per minute (0.0095 m^3/s) applied over the 1,500-square-foot (140 m^2) design area. Another example would be a manufacturing facility classified as ordinary hazard group 2 where a typical design area would be 1,500 square feet (140 m^2) and the design density would be 0.2 US gallons per minute (1.3×10^{-5} m^3/s) per 1 square foot (0.093 m^2) or a minimum of 300 US gallons per minute (0.019 m^3/s) applied over the 1,500-square-foot (140 m^2) design area.

After the design area and density have been determined, calculations are performed to prove that the system can deliver the required amount of water over the required design area. These calculations account for all of the pressure that is lost or gained between the water supply source and the sprinklers that would operate in the design area. This includes pressure losses due to friction inside the piping and losses or gains due to elevational differences between the source and the discharging sprinklers. Sometimes momentum pressure from water velocity inside the piping is also calculated.

Typically these calculations are performed using computer software but before the advent of computer systems these sometimes complicated calculations were performed by hand. This skill of calculating sprinkler systems by hand is still required training for a sprinkler system design technologist who seeks senior level certification from engineering certification organisations such as the National Institute for Certification in Engineering Technologies (NICET).

Sprinkler systems in residential structures are becoming more common as the cost of such systems becomes more practical and the benefits become more obvious. Residential sprinkler systems usually fall under a residential classification separate from the commercial classifications mentioned above. A commercial sprinkler system is designed to protect the structure and the occupants from a fire. Most residential sprinkler systems are primarily designed to suppress a fire in such a way to allow for the safe escape of the building occupants. While these systems will often also protect the structure from major fire damage, this is a secondary consideration. In residential structures sprinklers are often omitted from closets, bathrooms, balconies, garages and attics because a fire in these areas would not usually impact the occupant's escape route.

If water damage or water volume is of particular concern, a technique called *Water Mist Fire Suppression* may be an alternative. This technology has been under development for over 50 years. It hasn't entered general use, but is gaining some acceptance on ships and in a few residential applications. Mist suppression systems work by using the heat of the fire to 'flash' the water mist cloud to steam. This then smothers the fire. As such, mist systems tend to be highly effective where there is likely to be a free-burning hot fire. Where there is insufficient heat (as in a deep seated fire such as will be found in stored paper, no steam will be generated and the mist system will not extinguish the fire. Some tests have shown that the volume of water needed to extinguish a fire with such a system installed can be dramatically less than with a conventional sprinkler system.

Costs

In 2006, the hardware costs of sprinkler systems run from US$2 – $5 per square foot ($50/m^2), depending on type and location. However, speciality systems may cost as much as $10/square foot ($100/m^2). Systems can be installed during construction or retrofitted. Some communities have laws requiring residential sprinkler systems, where large municipal hydrant water supplies ("fire flows") are not available.

Nationwide in the United States, one and two-family homes generally do not require fire sprinkler systems, although the overwhelming loss of life due to fires occurs in these spaces. Residential sprinkler systems are inexpensive (about the same per square foot as carpeting or floor tiling), but require larger water supply piping than is normally installed in homes, so retrofitting is usually cost prohibitive.

According to the National Fire Protection Association (NFPA), fires in hotels with sprinklers averaged 78% less damage than fires in hotels without them (1983–1987). The NFPA says the average loss per fire in buildings with sprinklers was $2,300, compared to an average loss of $10,300 in unsprinklered buildings. The NFPA adds that there is no record of a fatality in a fully sprinklered building outside the point of fire origin. However, in a purely economic comparison, this is not a complete picture; the total costs of fitting, and the costs arising from non-fire triggered release must be factored.

The NFPA states that it "has no record of a fire killing more than two people in a completely sprinklered building where a sprinkler system was properly operating, except in an explosion or flash fire or where industrial fire brigade members or employees were killed during fire suppression operations."

The world's largest fire sprinkler manufacturer is the Simplex Grinnell division of Tyco International, other manufacturers / suppliers include The Viking Corporation, Victaulic, NNI Inc, P.u.P. Feuerschutz und Anlagenbau GmbH and Reliable Sprinkler Company.

Chapter 3

Pipe Fitting

Pipe fitting is the occupation of installing or repairing piping or tubing systems that convey liquid, gas, and occasionally solid materials. This work involves selecting and preparing pipe or tubing, joining it together by various means, and the location and repair of leaks.

Pipe fitting work is done in many different settings: HVAC, manufacturing, hydraulics, refineries, nuclear-powered Supercarriers and Fast Attack Submarines computer chip fab plants, power plant construction and other steam systems. Pipe fitters (sometimes called simply "fitters") are represented in the USA and Canada by the United Association of Journeymen and Apprentices of the Plumbing and Pipe Fitting Industry of the United States and Canada.

Figure: Soldered copper pipes

Fitters work with a variety of pipe and tubing materials including several types of steel, copper, iron, aluminium, and plastic. Pipe fitting is not plumbing; the two are related but separate trades. Pipe fitters who specialise in fire prevention are called Sprinklerfitters, another related, but separate trade.

Materials, techniques, and usages vary from country to country as different nations have different standards to install pipe.

Steel Pipe

Steel pipe (or black iron pipe) was once the most popular choice for supply of water and flammable gases. Steel pipe is still used in many homes and businesses to convey natural gas or propane fuel, and is a popular choice in fire sprinkler systems due to its high heat resistance. In commercial buildings, steel pipe is used to convey heating or cooling water to heat exchangers, air handlers, variable air volume (VAV) devices, or other HVAC equipment.

Steel pipe is sometimes joined using threaded connections, where tapered threads are cut into the end of the tubing segment, sealant is applied in the form of thread sealing compound or thread seal tape (also known as PTFE or Teflon tape), and it is then threaded into a corresponding threaded fitting using a pipe wrench. Beyond domestic or light commercial settings, steel pipe is often joined by welding, or by use of mechanical couplings made by companies such as Victaulic or Grinnell that hold the pipe joint together via a groove pressed or cut (a rarely used older practice), into the ends of the pipes.

Other variations of steel pipe include various stainless steel and chrome alloys. In high-pressure situations these are usually joined by TIG welding.

In Canada, with respect to natural gas (NG) and propane (LP gas), black iron pipe (BIP) is commonly used to connect an appliance to the supply. It must however be marked (either painted yellow or yellow banding attached at certain intervals) and certain restrictions apply to which nominal pipe size (NPS) can be put through walls and buildings. With propane in particular, BIP can be run from an exterior tank (or cylinder) provided it is well protected from the weather, and an anode-type of protection from corrosion is in place when the pipe is to be installed underground.

Copper Pipe

Copper tubing is most often used for supply of hot and cold water, and as refrigerant line in HVAC systems. There are two basic types of copper tubing, soft copper and rigid copper. Copper tubing is joined using flare connection, compression connection, or solder. Copper offers a high level of resistance to corrosion, but is becoming very costly.

Soft Copper

Soft (or ductile) copper tubing can be bent easily to travel around obstacles in the path of the tubing. While the work hardening of the drawing process used to size the tubing makes the copper hard/rigid, it is carefully annealed to make it soft again; it is therefore more expensive to produce than non-annealed, rigid copper tubing. It can be joined by any of the three methods used for rigid copper, and it is the only type of copper tubing suitable for flare connections. Soft copper is the most popular choice for refrigerant lines in *split-system* air conditioners and heat pumps.

Flare Connections

Flare connections require that the end of a tubing section be spread outward in a bell shape using a flare tool. A flare nut then compresses this bell-shaped end onto a male fitting. Flare connections are a labour intensive method of making connections, but are quite reliable over the course of many years.

Rigid Copper

Rigid copper is a popular choice for water lines. It is joined using a sweat, compression or crimped/pressed connection. Rigid copper, rigid due to the work hardening of the drawing process, cannot be bent and must use *elbow fittings* to go around corners or around obstacles. If heated and allowed to slowly cool, called annealing, then rigid copper will become soft and can be bent/formed without cracking.

Soldered Connections

Solder fittings are smooth, and easily slip onto the end of a tubing section. The joint is then heated using a torch, and solder is melted into the connection. When the solder cools, it forms a very strong bond which can last for decades. Solder-connected rigid copper is the most popular choice for water supply lines in modern buildings. In situations where many connections must be made at once (such as plumbing of a new building), solder offers much quicker and much less expensive joinery than compression or flare fittings. The term sweating is sometimes used to describe the process of soldering pipes.

Compression Connections

Compression fittings use a soft metal or thermoplastic ring (the compression ring or "olive") which is squeezed onto the pipe and into the fitting by a compression nut. The soft metal conforms to the surface of the tubing and the fitting, and creates a seal. Compression connections do not typically have the long life that sweat connections offer, but are advantageous in many cases because they are easy to make using basic tools. A disadvantage in compression connections is that they take longer to make than sweat, and sometimes require retightening over time to stop leaks.

Crimped or Pressed Connections

Crimped or pressed connections use special copper fittings which are permanently attached to rigid copper tubing with a powered crimper. The special fittings, manufactured with sealant already inside, slide over the tubing to be connected. Thousands of pounds-force per square inch of pressure are used to deform the fitting and compress the sealant against the inner copper tubing, creating a water tight seal. The advantages of this method are that it should last as long as the tubing, it takes less time to complete than other methods, it is cleaner in both appearance and the materials used to make the connection, and no open flame is used during the connection process. The disadvantages are that the fittings used are harder to find and cost significantly more than sweat type fittings.

Aluminium Pipe

Aluminium is sometimes used due to its low cost, resistance to corrosion and solvents, and its ductility. Aluminium tube is more desirable than steel for the conveyance of flammable solvents, since it cannot create sparks when manipulated. Aluminium tubing can be connected by flare or compression fittings, or it can be welded by the TIG or heliarc processes.

Glass Pipe

Tempered glass pipes are used for specialised applications, such as corrosive liquids, medical or laboratory wastes, or pharmaceutical manufacturing. Connections are generally made using specialised gasket or O-ring fittings.

Occupational Hazards

Pipe fitters are often exposed to hazardous or dangerous materials, such as asbestos, lead, ammonia, steam, flammable gases, various resins and solvents including benzene, and various refrigerants. Much progress was made in the 20th century toward eliminating or reducing hazardous materials exposures. Many aspects of hazardous materials are now regulated by law in most countries, including asbestos usage and removal, and refrigerant selection and handling.

Other occupational hazards include exposure to the weather, heavy lifting, crushing hazards, lacerations, and other risks normal to the construction industry.

Scope of Work

- Blueprint reading
- Detailing
- CAD Drawing coordinators
- Layout
- Supports and hanger Installation
- Pipe threading
- Pipe grinding
- Welding (MiG, GMAW, TiG, SMAW, Orbital)
- Plasma cutting
- Mechanical pipe cutting and grooving
- Gas arc cutting
- Rigging
- Brazing
- Soldering
- Mitering
- Tube bending
- Valve installation and repair
- Preparation and installation of medical gas piping.

Flange

A flange is an external or internal ridge, or rim (lip), for strength, as the flange of an iron beam such as an I-beam or a T-beam; or for attachment to another object, as the flange on the end of a pipe, steam cylinder, etc., or on the lens mount of a camera; or for a flange of a rail car or tram wheel. Thus flanged wheels are wheels with a flange on one side to keep the wheels from running off the rails. The term "flange" is also used for a kind of tool used to form flanges. Pipes with flanges can be assembled and disassembled easily.

Plumbing or Piping

A flange can also be a plate or ring to form a rim at the end of a pipe when fastened to the pipe (for example, a closet flange). A blind flange is a plate for covering or closing the end of a pipe. A flange joint is a connection of pipes, where the connecting pieces have flanges by which the parts are bolted together.

Figure: Surrey flange

Although the word flange generally refers to the actual raised rim or lip of a fitting, many flanged plumbing fittings are themselves known as 'flanges':

Common flanges used in plumbing are the Surrey flange or Danzey flange, York flange, Sussex flange and Essex flange. Surrey and York flanges fit to the top of the hot water tank allowing all the water to be taken without disturbance to the tank. They are often used to ensure an even flow of water to showers. An Essex flange requires a hole to be drilled in the side of the tank.

There is also a Warix flange which is the same as a York flange but the shower output is on the top of the flange and the vent on the side. The York and Warix flange have female adapters so that they fit onto a male tank, whereas the Surrey flange connects to a female tank.

Pipe Flanges

There are many different flange standards to be found worldwide. To allow easy functionality and inter-changeability, these are designed to have standardised dimensions. Common world standards include ASA/ANSI (USA), PN/DIN (European), BS10 (British/Australian), and JIS/KS (Japanese/Korean).

In most cases these are not interchangeable (e.g. an ANSI flange will not mate against a JIS flange). Further, many of the flanges in each standard are divided into "pressure classes", allowing flanges to be capable of taking different pressure ratings. Again these are not generally interchangeable (e.g. an ANSI 150 will not mate with an ANSI 300). These pressure classes also have differing pressure and temperature ratings for different materials. Unique pressure classes for piping can also be developed for a process plant or power generating station; these may be specific to the corporation, engineering procurement and construction (EPC) contractor, or the process plant owner.

The flange faces are also made to standardised dimensions and are typically "flat face", "raised face", "tongue and groove", or "ring joint" styles, although other obscure styles are possible.

Flange designs are available as "weld neck", "slip-on", "lap joint", "socket weld", "threaded", and also "blind".

ASME Standards (U.S.)

Figure: ASME type flange on a gas pipeline

Pipe flanges that are made to standards called out by ASME B16.5 or ASME B16.47 are typically made from forged materials and have machined surfaces. B16.5 refers to nominal pipe sizes (NPS) from ½" to 24". B16.47 covers NPSs from 26" to 60". Each specification further delineates flanges into pressure classes: 150, 300, 400, 600, 900, 1500 and 2500 psi for B16.5; B16.47 delineates its flanges into pressure classes 75, 150, 300, 400, 600, 900.

The gasket type and bolt type are generally specified by the standard(s); however, sometimes the standards refer to the ASME Boiler and Pressure Vessel Code (B&PVC) for details. These flanges are recognised by ASME Pipe Codes such as ASME B31.1 Power Piping, and ASME B31.3 Process Piping.

Materials for flanges are usually under ASME designation: SA-105 (Specification for Carbon Steel Forgings for Piping Applications), SA-266 (Specification for Carbon Steel Forgings for Pressure Vessel Components), or SA-182 (Specification for Forged or Rolled Alloy-Steel Pipe Flanges, Forged Fittings, and Valves and Parts for High-Temperature Service). In addition, there are many "industry standard" flanges that in some circumstance may be used on ASME work.

Other Countries

Flanges in other countries also are manufactured according to the standards for materials, pressure ratings, etc. Such standards include DIN, BS, and/or ISO standards.

Vacuum Flange

Figure: A KF-25 tee, o-ring, and clamp.

A vacuum flange is a flange at the end of a tube used to connect vacuum chambers, tubing and vacuum pumps to each other.

Vacuum Flange Types

Several vacuum flange standards exist, and the same flange types are called by different names by different manufacturers and standards organisations.

Figure: A CF (conflat) pipe and flange with copper gasket.

KF/QF

The ISO standard quick release flange is known by the names Quick Flange (QF), Klein Flange (KF) or NW, sometimes also as DN. The KF designation has been adopted by ISO, DIN, and Pneurop. KF flanges are made with a chamferred back surface that attached with a circular clamp and an elastomeric o-ring that is mounted in a metal centring ring. Standard sizes are indicated by the nominal inner diametre in millimetres for flanges 16 through 50 mm in diametre.

- DN16KF
- DN25KF
- DN40KF
- DN50KF

ISO

The ISO large flange standard is known as LF, LFB, MF or sometimes just ISO flange. As in KF-flanges, the flanges are joined by a centreing ring and an elastomeric o-ring. An extra spring-loaded circular clamp is often used around the large diametre o-rings to prevent them from rolling off from the centreing ring during mounting.

The ISO large flanges come in two varieties. The ISO-K (or ISO LF) flanges are joined with double claw clamps which clamp to a circular groove on the tubing side of the flange. The ISO-F (or ISO

LFB) flanges have holes for attaching the two flanges with bolts. Two tubes with ISO-K and ISO-F flanges can be joined together by clamping the ISO-K side with single claw clamps which are then bolted to the holes on the ISO-F side.

ISO large flanges are available in sizes from 63 to 500 mm nominal tube diametre).

- DN63LF (63.5 mm)
- DN100LF (102 mm)
- DN160LF (160 mm)
- DN200LF (200 mm)
- DN250LF (254 mm)
- DN320LF (316 mm)
- DN400LF (400 mm)
- DN500LF (500 mm)

CF (Conflat)

CF (ConFlat) flanges use a copper gasket and knife-edge flange to achieve an ultrahigh vacuum seal. The term "ConFlat" is a registered trademark of Varian, Inc., so "CF" is commonly used by other flange manufacturers. Each face of the two mating CF flanges has a knife edge which cuts into the softer metal gasket, providing an extremely leak-tight, metal-to-metal seal. Deformation of the metal gasket fills small defects in the flange, allowing Conflat flanges operate down to 10^{-13} torr (10^{-11} Pa) pressure. The gasket is partially recessed in a groove in each flange. The groove helps hold the gasket in place, which aligns the two flanges and also reduces gasket expansion during bake-out. For stainless steel conflat flanges baking temperatures of 450°C can be achieved; the temperature is limited by the choice of gasket material. CF flanges are sexless and interchangeable. North American flange sizes are given by flange outer diametre in inches while in Europe and Asia, sizes are given by tube inner diametre in millimetres:

European, Asian size	North American size [inches]
DN10	1
DN16	$1^{1}/_{3}$ ("mini")
DN25	$2^{1}/_{2}$
DN40 (or: DN35)	2¾

DN50	$3^{3}/_{2}$
DN63	4½
DN75	$4^{5}/_{2}$
DN100	6
DN125	6¾
DN160 (or: DN150)	8
DN200	10
DN250	12
	13¼
	14
	16½

Figure: A 60 kV high voltage electrical feedthrough on a 4-1/2 inch (or DN63) conflat flange

ConFlat gaskets were originally invented by William Wheeler and other engineers at Varian in an attempt to build a flange that would not leak after baking.

Wheeler

A Wheeler flange is a large wire seal flange often used on large vacuum chambers.

ASA

ANSI has a flange standard called ASA. These flanges are elastomeric o-ring seal and can be used for both vacuum and pressure applications. Flange sizes are indicated by tube nominal inner diametre or by flange outer diametre (in inches): 1 (4.25 O.D.), 1.5 (5.00 O.D.), 2 (6.00 O.D.), 3 (7.50 O.D.), 4 (9.00 O.D.), 6 (11.00 O.D.), 8 (13.5 O.D.), 10 (16.00 O.D.).

Vacuum Gaskets

To achieve a vacuum seal, a gasket is required. An elastomeric o-ring gasket can be made of Buna rubber, viton fluoropolymer, silicon rubber or teflon. O-rings can be placed in a groove or may be used in combination with a centreing ring or as a "captured" o-ring that is held in place by separate metal rings. Metal gaskets are used in ultra-high vacuum systems where the outgassing of the elastomer could be a significant gas load. A copper ring gasket is used with conflat flanges. Metal wire gaskets made of copper, gold or indium can be used.

Vacuum Feedthrough

A vacuum feedthrough is a flange that contains a vacuum-tight electrical, physical or mechanical connection to the vacuum chamber. An electrical feedthrough allows voltages to be applied to components under vacuum, for example a filament or heater. An example of a physical feedthrough is a vacuum tight connection for cooling water. A mechanical feedthrough is used for rotation and translation of components under vacuum. A wobble stick is a mechanical feedthrough device that can be used to pick up.

Flow Conditioning

Flow conditioning ensures that the "real world" environment closely resembles the "laboratory" environment for proper performance of inferential flowmetres like orifice, turbine, coriolis, ultrasonic etc.

Types of Flow

Basically, Flow in pipes can be classified as follows –

- Fully developed flow (found in world class flow laboratories)
- Pseudo fully developed flow
- Non-swirling, non-symmetrical flow
- Moderate swirling, non-symmetrical flow
- High swirling, symmetrical flow

Types of Flow Conditioners

Figure: showing The Flow conditioner

Flow conditioners shown in figure can be grouped into following three types –

- Those that eliminate swirl only (tube bundles)
- Those that eliminate swirl and non-symmetry, but do not produce pseudo fully developed flow
- Those that eliminate swirl, non-symmetry and pseudo fully developed flow (high performance flow conditioners).

Natural Gas Measurement

Natural gas that carries a lot of liquids with it is known as wet gas whereas natural gas that is produced without liquid is known dry gas. Dry gas is also treated as to remove all liquids. The effect of flow conditioning for various popular metres which is used in gas measurement is explained below.

Pipe Flow Conditions

The most important as well as most difficult to measure aspects of flow measurement are flow conditions within a pipe upstream of a metre. Flow conditions mainly refer to the gas velocity profile, irregularities in the profile, varying turbulence levels within the velocity or turbulence intensity profile, swirl and any other fluid flow characteristics which will cause the metre to register flow different than that expected. It will change the value from the original calibration state referred to as reference conditions that are free of installation effects.

Installation Effects

Installation effects such as insufficient straight pipe, exceptional pipe roughness or smoothness, elbows, valves, tees and reducers causes the flow conditions within the pipe to vary from the reference conditions. How these installation effects impact the metre is very important since devices which create upstream installation effects are common components of any standard metring design. Flow Conditioning refers to the process of artificially generating a reference, fully developed flow profile and is essential to enable accurate measurement while maintaining a cost-competitive metre standard design. The metre calibration factors are valid only of geometric and dynamic similarity exists between the metring and calibration conditions. In fluid mechanics, this is commonly referred to as the Law of Similarity.

Law of Similarity

The principle of Law of Similarity is used extensively for theoretical and experimental fluid machines. With respect to calibration of flowmetres, the Law of Similarity is the foundation for flow measurement standards. To satisfy the Law of Similarity, the central facility concept requires geometric and dynamic similarity between the laboratory metre and the installed conditions of this same metre over the entire custody transfer period.

This approach assumes that the selected technology does not exhibit any significant sensitivity to operating or mechanical variations between calibrations. The metre factor determined at the time of calibration is valid if both dynamic and geometric similarity exists between the field installation and the laboratory installation of the artifact.

A proper manufacturer's experimental pattern locates sensitive regions to explore, measure and empirically adjust. The manufacturer's recommended correlation method is a rational basis for performance prediction provided the physics do not change. For instance, the physics are different between subsonic and sonic flow. To satisfy the Law of Similarity the in situ calibration concept requires geometric and dynamic similarity between the calibrated metre and the installed conditions of this same metre over the entire custody transfer period. This approach assumes that the selected technology does not exhibit any significant sensitivity to operating or mechanical variations between calibrations. The metre factor determined at the time of calibration is valid if both dynamic and geometric similarity exists in the "field metre installation" over the entire custody transfer period.

Velocity Flow Profile

The most commonly used description of flow conditions within the pipe is the velocity flow profile. The shape of the velocity flow profile is given by the following equation,

$$\frac{U_y}{U_{max}} = \left[1 - \frac{Y}{R}\right]^{1/n} \tag{1}$$

The value of n determines the shape of the velocity flow profile. The eq.(1) can be used to determine the flow profile's shape within the pipe by fitting a curve to experimentally measured velocity data. In 1993, the transverse velocities were being measured within the high pressure natural gas environment using hot wire technology to accomplish the data fit. A fully developed flow profile was used as the reference state for metre calibration and determination of Coefficient of Discharge (Cd). For Reynolds Number 10^5 to 10^6 n is approximately 7.5; for Re of 10^6, n is approximately 10.0 where a fully developed profile in a smooth pipe was assumed. Since n is a function of Reynolds Number and friction factor, more accurate values of n can be estimated by using the eq.(2),

$$n = \frac{1}{\sqrt{f}}. \tag{2}$$

Where, f is friction factor. A good estimate of a fully developed velocity flow profile can be used for those without adequate equipment to actually measure the velocities within the pipe. The following straight pipe equivalent length in eq.(3) was utilised to ensure a fully developed flow profile exists.

$$PipeDiameters = 4.4D\left[R_e\right]^{1/6} \tag{3}$$

In eq.(3) the pipe lengths required is significant, hence we need some devices that can able to condition the flow over a shorter pipe length allowing metring packages to be cost competitive and accurate. Here the velocity flow profile is generally three-dimensional. Normally the description requires no axial orientation indication if the profile is asymmetric and if it does exists, then axial orientation with respect to some suitable plane of reference is required. Asymmetry exists downstream of installation effects such as elbows or tees. Usually, the velocity flow profile is described on two planes 90° apart. Using the latest software technology a full pipe cross sectional description of the velocity profile is possible provided sufficient data points are given.

Turbulence Intensity

The second description of the flow field state within the pipe is the turbulence intensity. According to an experiment in 1994, the metring errors may exist even when the velocity flow profile is fully developed with perfect pipe flow conditions. Conversely, it was found zero metring error at times when the velocity profile was not fully developed. Hence this beahviour was referred to the turbulence intensity of the gas flow that can cause metring bias error. This beahviour accounts in part for the less than adequate performance of the conventional tube bundle.

Swirl

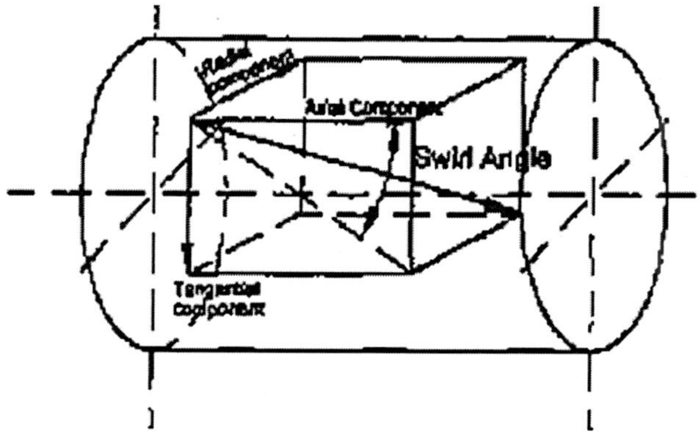

Figure: showing Swirl angle

The third description of the flow field's state is swirl. Swirl is the tangential flow component of the velocity vector. The velocity profile should be referred to as the axial velocity profile. As the velocity vector can be resolved into three mutually orthogonal components, the velocity profile only represents the axial component of velocity. In real pipeline conditions, such as downstream of elbows two or more mechanisms of swirl may be present.

Effects of Flow Conditioning on Orifice Metre

The basic orifice mass flow equation provided by API 14.3 and ISO 5167 is given as,

$$q_m = (C_d)(E_v)(Y)\left[\frac{\pi}{4}\right](d)^2\sqrt{2\rho\Delta P} \qquad (4)$$

Where, q_m = Mass flow

C_d = Coefficient of discharge

E_v = Velocity of approach factor

Y = Expansion factor

d = orifice diametre

p = density of the fluid

ΔP = differential pressure

Now to use the eq.(4), the flow field entering the orifice plate must be free of swirl and exhibit a fully developed flow profile. API 14.3 (1990) and ISO standards determined the Coefficient of Discharge by completing numerous calibration tests where the indicated mass flow was compared to the actual mass flow to determine coefficient of discharge.

In all testing the common requirement was a fully developed flow profile entering the orifice plate. Accurate standard compliant metre designs must therefore ensure that a swirl free, fully developed flow profile is impinging on the orifice plate. There are numerous methods available to accomplish this. These methods are commonly known as "flow conditioning".

The first installation option is to revert to no flow conditioning, but adequate pipe lengths must be provided by the eq.(2) mentioned above. This generally makes the manufacturing costs for a flow measurement facility unrealistic due to excessively long metre tubes; Imagine metre tubes 75 diametres long.

The second and most well known option is the 19-tube tube-bundle flow conditioner. The majority of flow installations in North America contain the tube bundle. With the help of hot wire, pitot tube and laser-based computerised measurement systems which allow detailed measurement of velocity profile and turbulence intensity; we know that the tube bundle does not provide fully developed flow. Therefore, this device is causing biased orifice flow measurement. As a result of these recent findings, few tube bundles are specified for flow measurement and reduce the use of such device. Numerous references are available providing performance results indicating less than acceptable metre performance when using the conventional 19-tube test bundle. The individual results should be reviewed to ascertain details such as beta ratio, metre tube lengths, Re and test conditions.

The general indications are that the conventional tube bundle will cause the orifice installation to over register flow values up to 1.5% when the tube bundle is 1 pipe diametre to approximately 11 pipe diametres from the orifice plate.

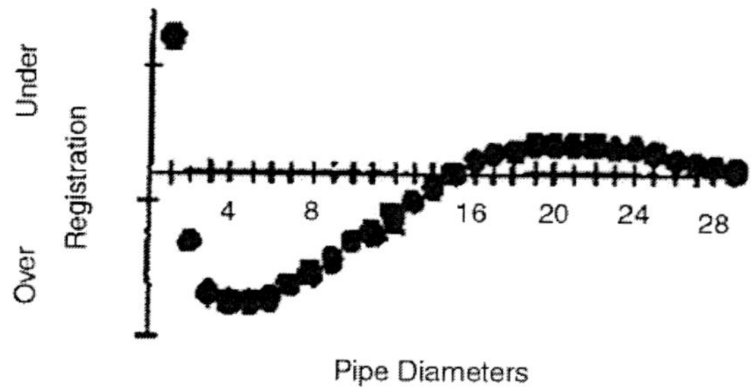

Figure: showing Conventional tube bundle performance

This is caused by a flat velocity profile that creates higher differential pressures than with a fully developed profile. There is a crossover region from approximately 10 to 15 pipe diametres where the error band is approximately zero. Then a slight under-registration of flows occurs for distances between approximately 15 to 25 pipe diametres. This is due to a peaked velocity profile that creates lower differential pressures than a fully developed profile. At distances greater than 25 pipe diametres the error asymptotes to zero. An additional drawback of the conventional 19 tube, tube bundle is variation in sizing.

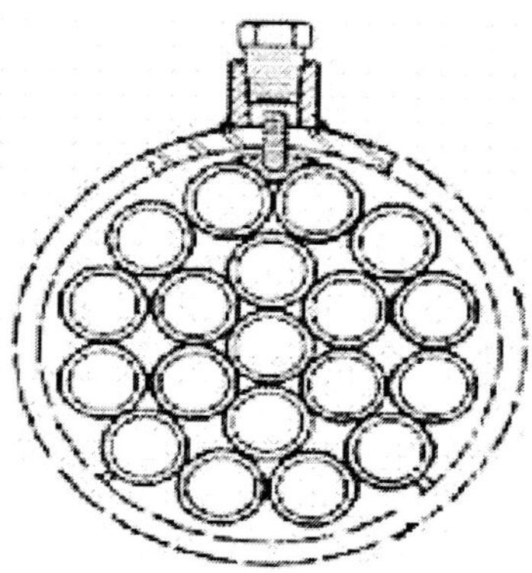

Figure: showing the 19-tube bundle

The conventional tube bundle provides errors very much dependent on installation details, that is, the elbows on and out of plane, tees, valves and distances from the last pipe installation to the conditioner and conditioner to the orifice plate. These errors have a great significance. Therefore the latest findings regarding conventional tube bundle performance should be reviewed prior to metre station design and installation. The final installation option for orifice metring is perforated plate flow conditioners. There is a variety of perforated plates have entered the market. These devices generally are designed to rectify the drawbacks of the conventional tube bundle (accuracy and repeatability insufficiency). The reader is cautioned to review the performance of the chosen perforated plate carefully prior to installation. A flow conditioner performance test guideline should be utilised to determine performance. The key elements of a flow conditioner test are -

1. Perform a baseline calibration test with an upstream length of 70 to 100 pipe diametres of straight metre tube. The baseline Coefficient of Discharge values should be within the 95% confidence interval for the RG orifice equation (i.e. the coefficient of discharge equation as provided by AGA-3).
2. Select values of upstream metre tube length, and flow conditioner location, to be used for the performance evaluation. Install the flow conditioner at the desired location. First, perform a test for either the two 90° elbows out-of-plane installation, or the high swirl installation for $\beta = 0.40$ and for $\beta = 0.67$. This test will show whether the flow conditioner removes swirl from the disturbed flow. If the ΔCd is within the acceptable region for both values of β i.e. 0.40 and 0.67, and if the Cd results vary as $(\beta)^{3.5}$, then the conditioner is successful in removing swirl. The tests for the other three installations namely, good flow conditions, partly closed valve and highly disturbed flow) may be performed for $\beta = 0.67$, and the results for other (i ratios predicted from the $\Delta Cd - (\beta)^{3.5}$ correlation. Otherwise, the tests should be performed for a range of p ratios between 0.20 and 0.75.
3. Perform test and determine the flow conditioner performance for the flow conditioner installed in good flow conditions, downstream of a half closed valve, and for either the double 90° elbow out-of-plane or the high swirl installation.

Effects of Flow Conditioning on Turbine Metre

The turbine metre is available in various manufacturer's configurations of a common theme; turbine blades and rotor configured devices. These devices are designed such that when a gas stream passes through them they will spin proportionally to the amount of gas passing over the blades in a repeatable fashion. Accuracy is then ensured by completion of a calibration, indicating the relationship between rotational speed and volume, at various Reynolds Numbers.

The fundamental difference between the orifice metre and the turbine metre is the flow equation derivation. The orifice metre flow calculation is based on fluid flow fundamentals (a 1st Law of Thermodynamics derivation utilising the pipe diametre and vena contracta diametres for the continuity equation). Deviations from theoretical expectation can be assumed under the Coefficient of Discharge. Thus, one can manufacture an orifice metre of known uncertainty with only the measurement standard in hand and access to a machine shop.

Conversely, the turbine metre operation is not rooted deeply in fundamentals of thermodynamics. This is not to say that the turbine metre is in any way an inferior device. There are sound engineering principles providing theoretical background. It is essentially an extremely repeatable device that is then assured accuracy via calibration.

The calibration provides the accuracy. It is carried out in good flow conditions (flow conditions free of swirl and a uniform velocity flow profile) this is carried out for every metre manufactured. Deviations from the as-calibrated conditions would be considered installation effects, and the sensitivity of the turbine metre to these installation effects is of interest. The need for flow conditioning is driven from the sensitivity of the metre to deviations from as calibrated conditions of swirl and velocity profile. Generally, recent research indicates that turbine metres are sensitive to swirl but not to the shape of the velocity profile. A uniform velocity profile is recommended, but no strict requirements for fully developed flow profiles are indicated. Also, no significant errors are evident when installing single or dual rotor turbine metres downstream of two elbows out-of-plane without flow conditioning devices.

Effects of Flow Conditioning on Ultrasonic Metre

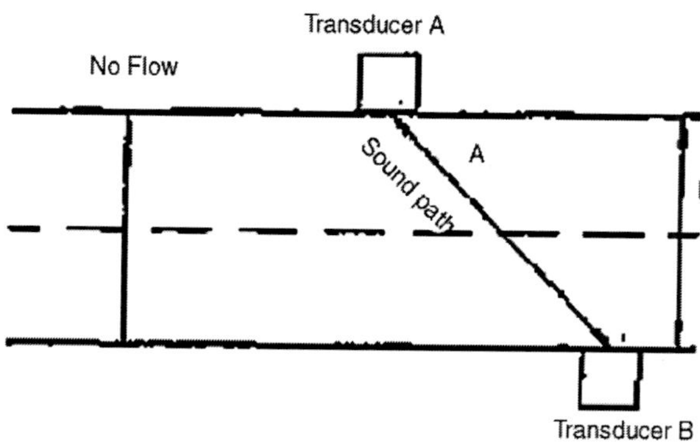

Figure: showing ultrasonic metre sound path-no flow

Due to the relative age of the technology, it may be beneficial to discuss the operation of the multipath ultrasonic metre to illustrate the effects of flow profile distortion and swirl. There are various types of flow measurements utilising high frequency sound. The custody transfer measurement devices available today utilise the time of travel concept. The difference in time of flight with the flow is compared to the time of flight against the flow. This difference is used to infer average flow velocity on the sound path.

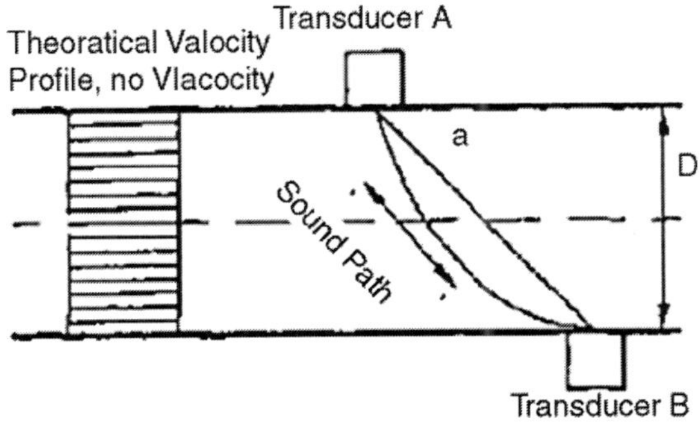

Figure: showing Ultrasonic metre sound path-uniform velocity profile

The resulting flow equation for the mean velocity experienced by the sound path is given by,

$$\overline{V}_{flow} = \left[\frac{1}{T_{ab}} - \frac{1}{T_{ba}}\right]\left[\frac{Dist_{Soundpath}}{2\cos\phi}\right] \tag{5}$$

The case of no flow gives the actual path of the sound when there is zero flow (by equating eq.(5) to zero). In case of theoretical flow profile, say a uniform velocity flow profile where the no-slip condition on the pipe walls is not applied.

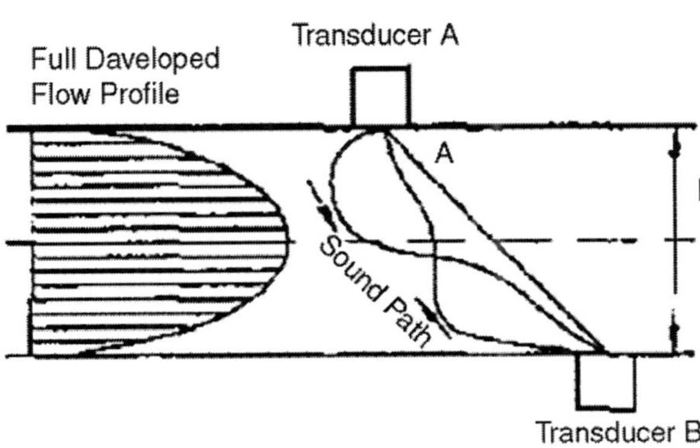

Figure: showing Ultrasonic metre sound path-fully developed flow

A theoretical derivation of the Mean velocity equation for this sound path becomes much more complicated. In case of a perfect fully developed real velocity profile of Ultrasonic metre indicating a possible sound path as a result of an installation in a real flow.

Here a mathematical derivation for this Ultrasonic metre is also becomes very complicated. Developing a robust flow algorithm to calculate the mean flow velocity for the sound path can be quite complicated. Now add to this; sound path reflection from the pipe wall, multipaths to add degrees of freedom, swirl and departure from axisymmetric fully developed flow profile and the problem of integrating the actual velocity flow profile to yield volume flow rate can be an accomplishment. Hence the real performance of ultrasonic metres downstream of perturbations, and the need for calibrations is required.

Effects of Flow Conditioning on Coriolis Metre

It poses a complex fluid structure interaction problem in case of two-phase operation. There is a scarcity of theoretical models available to predict the errors reported by Coriolis metre in aforementioned conditions.

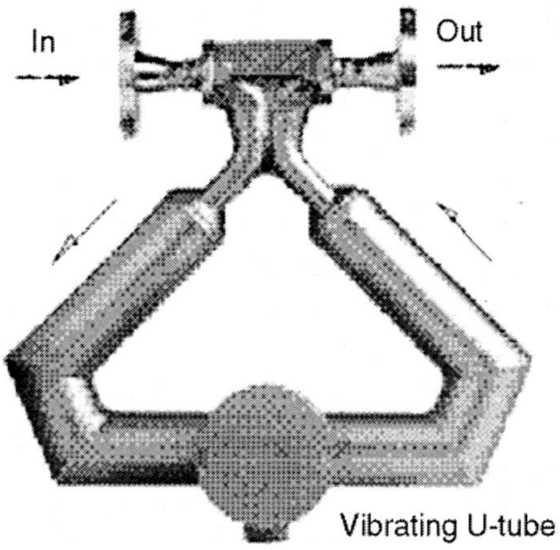

Figure: showing Coriolis metre

Flow conditioners make no effect on metre accuracy while using wet gas due to the annular flow regime, which is not highly affected by flow conditioners. In single-phase conditions, Coriolis metre gives accurate measurement even in presence of severe flow disturbances. There is no need for flow conditioning before the metre to obtain accurate readings from it, which would be the case in other metring technologies like orifice and turbine. On the other hand in two-phase flows, the metre consistently gives negative errors. The use of flow conditioners clearly affects the reading of the metre in aerated liquids. This phenomenon can be used to get fairly accurate estimate of flow rate in low gas volume fraction liquid flows.

Fluid Flow Measurement

Flow conditioning makes a huge effect on the accuracy of liquid turbine metre which results into flow disturbances. These effects are mainly caused by debris on strainer screens, for various upstream piping geometries and different types of flow conditioners. The effectiveness of a flow conditioner can be indicated by the following two key measurements:

- Percentage variation of an average metre factor over the defined range of flow disturbances for a given flow rate and inlet piping geometry. The lesser the value of percentage variation of an average metre factor over the range of flow disturbances, the better will be the performance of flow conditioner.
- Percentage metre factor repeatability for each flow disturbance, at a given flow rate and inlet piping geometry. The lesser the value of percentage metre factor repeatability at a given set of installation/operating conditions, the better will be the performance of flow conditioner.

Friction Loss

Friction loss refers to that portion of pressure lost by fluids while moving through a pipe, hose, or other limited space. In mechanical systems such as internal combustion engines, it refers to the power lost overcoming the friction between two moving surfaces.

Gooseneck (Piping)

A gooseneck (or goose neck) is a 180° pipe fitting at the top of a vertical pipe that prevents entry of water. Common implementations of goosenecks are ventilator piping or ducting for bathroom and kitchen exhaust fans, ship holds, landfill methane vent pipes, or any other piping implementation exposed to the weather where water ingress would be undesired.Gooseneck may also refer to a style of kitchen or bathroom faucet with a long vertical pipe terminating in a 180° bend.

The name gooseneck comes from the similarity of the pipe fitting to the bend in a goose's neck.

To avoid hydrocarbon accumulation, a thermosiphon should be installed at the low point of the gooseneck.

Collapsible Flow

Collapsible flow is a phenomenon that occurs in steady flow in tubes with significant distensibility, or the capability of swelling or stretching, under conditions of lower internal pressure relative to pressure outside the tube. Such conditions occur rarely in industrial applications but are very common in biological studies such as blood flow in veins and air flow in lungs.

When a flow is driven through a deformable channel or tube, interactions between fluid-mechanical and elastic forces can lead to a variety of biologically significant phenomena, including nonlinear pressure-drop/flow-rate relations, wave propagation, and the generation of instabilities. Understanding the physical origin and nature of these phenomena remains a significant experimental, analytical, and computational challenge, involving unsteady flows at low or high Reynolds numbers, large-amplitude fluid-structure interactions, free-surface flows, and intrinsically 2D or 3D motion. Whereas frequently the internal flow involves a single fluid phase (albeit often of a complex biological fluid such as blood), in many instances the presence of two or more distinct flowing phases is of primary importance (as is the case for air-liquid flows in peripheral lung airways, for example).

Single Phase in Collapsible Tubes

Venous collapse is important during exercise, when muscular compression of leg veins is used to pump blood against gravity up to the heart, and in therapeutic compression of leg veins for the treatment of deep-vein thrombosis

partial vessel collapse occurs in vessels which undergo conditions of higher external pressure relative to the fluid within and can be difficult to predict mathematically. As such, devices such as a Starling Resistor are often used to predict fluid flow under these conditions.

Fluid is forced through an elastically deforming tube which passes through a region of high external pressure causing a flattening of the tube depending on the relative pressures of the inside and outside of the tube. In the absence of any flow (puD pd), an increase in pe generates a compressive stress in the tube wall causing it to buckle from a circular to an elliptic cross-section (except, of course, near its ends, where it is attached to the rigid tubes). Buckling to a shape with more than two lobes may arise in short, tethered, or inhomogeneous tubes. Once buckled, the tube becomes highly compliant so that small additional increases in pe lead to a substantial reduction in cross-sectional area ®. Further compression leads to contact of the opposite tube walls, first at a point, and then along a line; once in opposite-wall contact, the tube's compliance falls because strong bending forces in the tube wall at the bulbous end of each lobe provide an increasing resistance to area reductions. The "tube law," the relation between transmural pressure P-Pexternal (where p is the internal pressure) and, for a long thin-walled tube can be approximated by thin-shell theory for an axially uniform elastic ring.

Cured-in-Place Pipe

A cured-in-place pipe (CIPP) is one of several trenchless rehabilitation methods used to repair existing pipelines. CIPP is a jointless, seamless, pipe-within-a-pipe with the capability to rehabilitate pipes ranging in diametre from 0.1–2.8 metre (4"–110"). As one of the most widely used rehabilitation methods CIPP has application in water, sewer, gas, and chemical pipelines.

The CIPP Process

A resin-saturated felt tube made of polyester is inverted or pulled into a damaged pipe. It is usually done from the upstream access point (manhole or excavation). It is possible to invert the liner upstream (e.g. from the downstream access point) but this carries greater risk. It is possible to install a liner from the downstream access point, upstream to a blind end however this carried the highest risk of all the CIPP installation methodologies. Little to no digging is involved in this trenchless process, making for a more environmentally friendly method than traditional "dig and replace" pipe repair methods. The liner can be inverted using water or air pressure.

The pressure required for inversion can be generated using pressure vessels, scaffolds or a "Chip unit". Hot water, UV light, ambient cured or steam is used to cure the resin and form a tight-fitting, jointless and corrosion-resistant replacement pipe. Service laterals are restored internally with robotically controlled cutting devices. These can be sealed with additional seals into the lateral connection (Lateral Junction Repair). The rehabilitated pipe is then inspected by closed-circuit television (CCTV). CIPP is considered a trenchless technology.

History

In 1971, Eric Wood implemented the first cured-in-place pipe technology in London, England. He called the CIPP process insitu form, derived from the Latin meaning "form in place." Wood applied for U.S. patent no. 4009063 on January 29, 1975. The patent was granted February 22, 1977 and was commercialised by Insituform Technologies until it entered the public domain on February 22, 1994.

Advantages

As a trenchless technology, CIPP does not require excavation to rehabilitate a pipeline that is either leaking or structurally unsound. Depending upon design considerations an excavation may be made, but the liner is often installed through a manhole or other existing access point. Anything larger than 60" must be excavated in order to install. Liner is installed as it is wet out on site in these instances. In the case of sewerlines, lateral connections are also restored without excavation via a remote controlled device that drills a hole in the liner at the point of the lateral connection.

If larger than 24" and it is safe to do so someone will reinstate laterals by hand. CIPP has a smooth interior and no joints. While CIPP can repair a pipe with bends, special design considerations must be taken into account to prevent wrinkling and stretching. CIPP can effectively reduce infiltration and leaks in pipeline systems without digging.

Disadvantages and Limitations

Except for very common sizes, liners are not usually stocked and must be made specifically for each project. CIPP requires bypass of the flow in the existing pipeline while the liner is being installed. The curing may take from five hours to 30 hours depending on pipe diametre and must be carefully monitored, inspected, and tested. Obstructions in the existing pipeline, such as protruding laterals, must be removed prior to installation.

Cost should be compared with similar methods such as Shotcrete, thermoformed pipe, close-fit pipe, spiral wound pipe and sliplining as these other methods can provide a similar design solution for similar or less cost in certain situations. CIPP must also be carefully monitored for release of chemical agents utilised in the reaction process for contamination downstream of rehabilitated pipes.

The liner material used for common sizes is normally a felted fabric (weave) and does not go around bends well without wrinkling and going out of round on corners. Liners used for pipes with bends (particularly 100 mm pipes) are made from a woven fabric allowing it to go around bends with minimal wrinkling. The more flexible the liner, the more care needs to be taken during inversion to ensure the liner remains on the correct alignment.

Corrugated Stainless Steel Tubing

Corrugated Stainless Steel Tubing (CSST) is a type of conduit used for natural gas heating in homes. It was introduced in the United States in 1988. CSST consists of a continuous, flexible stainless-steel pipe with an exterior PVC covering. The piping is produced in coils that are air-tested for leaks.

In the United States, about 500,000 new homes per year have CSST installed; that number is about half of the number of new homes that have natural gas for heating or cooking. As of 2007, about 2 million homes in the United States had CSST installed.

CSST is installed much like electrical wiring, and thus more quickly than traditional black-steel pipe, but the cost of the material is several times more than black-steel pipe.

In mid-2007, four CSST makers settled a class-action lawsuit. The settlement provides money to fix problem pipes The settlement provided money to fix problem pipes by bonding the CSST to the electrical systems with thicker wire.

Hazen–Williams Equation

The Hazen–Williams equation is an empirical formula which relates the flow of water in a pipe with the physical properties of the pipe and the pressure drop caused by friction. It is used in the design of water pipe systems such as fire sprinkler systems, water supply networks, and irrigation systems. It is named after Allen Hazen and Gardner Stewart Williams.

The Hazen–Williams equation has the advantage that the coefficient C is not a function of the Reynolds number, but it has the disadvantage that it is only valid for water. Also, it does not account for the temperature or viscosity of the water.

General Form

The general form of the equation relates the mean velocity of water in a pipe with the geometric properties of the pipe and slope of the energy line.

$$V = kCR^{0.63}S^{0.54}$$

where:

- V is velocity
- k is a conversion factor for the unit system (k = 1.318 for US customary units, k = 0.849 for SI units)

- C is a roughness coefficient
- R is the hydraulic radius
- S is the slope of the energy line (head loss per length of pipe or h_f/L)

Typical *C* factors used in design, which take into account some increase in roughness as pipe ages are as follows:

Material	*C Factor low*	*C Factor high*
Asbestos-cement	140	140
Cast iron	100	140
Cement-Mortar Lined Ductile Iron Pipe	140	140
Concrete	100	140
Copper	130	140
Steel	90	110
Galvanized iron	120	120
Polyethylene	140	140
Polyvinyl chloride (PVC)	130	130
Fibre-reinforced plastic (FRP)	150	150

Pipe Equation

The general form can be specialised for full pipe flows. Taking the general form

$$V = kCR^{0.63}S^{0.54}$$

and exponentiating $1/0.54$ each side by gives (rounding exponents to 2 decimals)

$$V^{1.85} = k^{1.85}C^{1.85}R^{1.17}S$$

Rearranging gives

$$S = \frac{V^{1.85}}{k^{1.85}C^{1.85}R^{1.17}}$$

The flow rate *Q* = *V A*, so

$$S = \frac{V^{1.85}A^{1.85}}{k^{1.85}C^{1.85}R^{1.17}A^{1.85}} = \frac{Q^{1.85}}{k^{1.85}C^{1.85}R^{1.17}A^{1.85}}$$

The hydraulic radius *R* (which is different from the geometric radius *r*) for a full pipe of geometric diametre *d* is *d*/4; the pipe's cross sectional area *A* is $\pi d^2/4$, so

$$S = \frac{4^{1.17}4^{1.85}Q^{1.85}}{\pi^{1.85}k^{1.85}C^{1.85}d^{1.17}d^{3.70}} = \frac{4^{3.02}Q^{1.85}}{\pi^{1.85}k^{1.85}C^{1.85}d^{4.87}} = \frac{4^{3.02}}{\pi^{1.85}k^{1.85}}\frac{Q^{1.85}}{C^{1.85}d^{4.87}} = \frac{7.916}{k^{1.85}}\frac{Q^{1.85}}{C^{1.85}d^{4.87}}$$

U.S. customary Units (Imperial)

When used to calculate the pressure drop using the US customary units system, the equation is:

$$P_d = \frac{4.52 \quad L \quad Q^{1.85}}{C^{1.85} \quad d^{4.87}}$$

where:

P_d = pressure drop over a length of pipe, psig (pounds per square inch gauge pressure)

L = length of pipe, ft (feet)

Q = flow, gpm (gallons per minute)

d = inside pipe diametre, in (inches)

SI Units

When used to calculate the pressure drop with the International System of Units, the equation becomes:

$$h_f = \frac{10.67 \quad L \quad Q^{1.85}}{C^{1.85} \quad d^{4.87}}$$

where:

- h_f = pressure loss over a length of pipe, m (head pressure)
- L = length of pipe, m (metres)
- Q = volumetric flow rate, m^3/s (cubic metres per second)
- d = inside pipe diametre, m (metres)

Heat-shrinkable Sleeve

Heat-shrinkable sleeve (or commonly "shrink sleeve") is a corrosion protective coating for pipelines in the form of a wraparound or tubular sleeve that is field-applied.

History

The first heat-shrinkable sleeves were introduced over 35 years ago when polyethylene pipeline coatings started to replace bituminous or tape coatings in the oil and gas industry. Back then, the processing for polyethylene to make the sleeve backing was new technology and the adhesives used in sleeves were much the same as those used on pipeline coatings.

The technology used to make sleeves has advanced significantly since then, with new methods of cross-linking the polyolefin backings and new generation adhesives that are formulated to provide performance under more demanding pipeline conditions.

Composition & Manufacture

Polyolefin Backing

Heat-shrinkable means just that, heat them up and they shrink, or more correctly, they recover in length. A heat-shrinkable sleeve starts out with a thick extruded polyolefin sheet (polyethylene or polypropylene) that is formulated to be cross-linkable. After extruding the thick sheet, it is taken to the "beam" where it is passed under a unit that subjects the sheet to electron irradiation. The irradiation process cross-links the polyolefin. This improves the molecular structure such that the polyolefin will work as part of a heat-shrinkable sleeve and provide the required level of mechanical protection while in-service. It makes the polyolefin perform more like a tough, heat-resistant, elastic material or rubber, rather than like a plastic material.

After cross-linking, the sheet is stretched by feeding it into a machine that heats it up, stretches it and cools it down. Because the sheet has been cross-linked, after stretching, it will want to recover to its original length when re-heated.

In recent years, many manufacturers had already developed their technologies of extruding and expansion of polyolefin backing. In the past, the production process of backing was done by extruding, cross-linking and expansion. However, in order to increase the production efficiency, some of manufacturers expand the backing during extruding, and then send the backing to e-beam for the cross-linking process.

Adhesives and Functions

An adhesive is then applied to the sheet and various manufacturers use proprietary techniques depending on the type, viscosity and melting temperature of the adhesive. The adhesive is the key to ultimate performance of the installed system, which is why different adhesive types will be specified depending on the pipeline operating conditions.

The adhesive has many functions; it adheres the installed sleeve to the steel at the coating cutback and mainline coating, it resists shear forces imparted by soil pressure after the pipeline is buried and provides long term corrosion protection to the steel. The choice of which adhesive to use is based on the pipeline design and operating conditions. As an example, for small diametre flow lines operating at ambient temperatures, a soft mastic-based adhesive may be chosen, while on large diametre pipelines operating at higher temperatures, a hard, semi-crystalline hot-melt adhesive is used. The adhesive needs to be chosen based on its corrosion protection properties, adhesion strength, and resistance to shear forces imparted by pipe movement and the effects of soil pressures. The coated sheet is then cut into individual sleeves suitable for application on a pipeline. As mentioned before, the sheet is stretched and wants to recover when heated, so a sealing strip or "closure" is applied during sleeve installation so that the sleeve will stay in place during and after recovery.

Epoxy Primer

A final component is an optional epoxy primer. Primers for heat-shrinkable sleeves work in the same manner as an FBE primer does when it is specified on 3-layer polyolefin pipeline coatings and is typically applied between 150 μm and 300 μm thick. Usually, the primer of heat shrinkable sleeve is two components non-solvent Epoxy, one is primer base and the other is curing agent.

Use

When steel pipelines are built, they commonly consist of 10~12m long sections of steel pipe that has had a corrosion protective coating applied to it in a factory. The factory will leave an uncoated area at each end of the pipe called a "cutback" so that when welding the pipe sections together, the coating is not damaged. Heat-shrinkable sleeves are applied onto the cutback at the field weld or "field joint" during the construction of a pipeline.

As described above, the heat-shrinkable sleeves have an adhesive that sticks the sleeve to the cutback and the factory applied mainline coating and also acts as a corrosion protective layer. The backing provides mechanical protection against abrasion and soil stress forces after the pipeline is buried.

Heat wrap tape may used in addition for pipe bends, or as an alternative method for wrapping the whole pipe.

Companies Offering Heat-shrinkable Sleeves

- CCL Label Decorative Sleeves
- Raychem
- Canusa-CPS
- Denso,
- TIAL,
- Dasheng CJDS
- Stepko

Main Standards and Certificates

- DVGW,
- ISO 21809-3,
- EN 12068,
- DIN 30678,
- NACE SP 0303,
- Shell standard.

Hot Tapping

Hot tapping, or pressure tapping, is the method of making a connection to existing piping or pressure vessels without the interruption of emptying that section of pipe or vessel. This means that a pipe or tank can continue to be in operation whilst maintenance or modifications are being done to it. The process is also used to drain off pressurised casing fluids.

Hot tapping is also the first procedure in line stopping, where a hole saw is used to make an opening in the pipe, so a line plugging head can be inserted.

Situations in which welding operations are prohibited on equipment which contains:

- Mixtures of gases or vapours within their flammable range or which may become flammable as a result of heat input in welding operations.
- Substances which may undergo reaction or decomposition leading to a dangerous increase in pressure, explosion or attack on metal. In this context, attention is drawn to the possibility that under certain combinations of concentration, temperature and pressure, acetylene, ethylene and other unsaturated hydrocarbons may decompose explosively, initiated by a welding hot spot.
- Oxygen-enriched atmospheres in the presence of hydrocarbons which may be present either in the atmosphere or deposited on the inside surface of the equipment or pipe.
- Compressed air in the presence of hydrocarbons which may be present either in the air or deposited on the inside surfaces of the equipment or pipe.
- Gaseous mixtures in which the partial pressure of hydrogen exceeds 700 kPa gauge, except where evidence from tests has demonstrated that hot-tapping can be done safely.

Based on the above, welding on equipment or pipe which contains hazardous substances or conditions as listed below (even in small quantities) shall not be performed unless positive evidence has been obtained that welding/hot tapping can be applied safely.

Hydrostatic Test

A hydrostatic test is a way in which leaks can be found in pressure vessels such as pipelines and plumbing. The test involves placing water, which is often dyed for visibility, in the pipe or vessel at the required pressure to ensure that it will not leak or be damaged. It is the most common method employed for testing pipes and vessels. Using this test helps maintain safety standards and durability of a vessel over time. Newly manufactured pieces are initially qualified using the hydrostatic test. They are then continually re-qualified at regular intervals using the *proof pressure test* which is also called the *modified hydrostatic test.* Hydrostatic testing is also a way in which a gas pressure vessel, such as a gas cylinder or a boiler, is checked for leaks or flaws. Testing is very important because such containers can explode if they fail when containing compressed gas.

Testing Procedures

Hydrostatic tests are conducted under the constraints of either the industry's or the customer's specifications. The vessel is filled with a nearly incompressible liquid-usually water or oil-and examined for leaks or permanent changes in shape. The test pressure is always considerably higher than the operating pressure to give a margin for safety. This margin of safety is typically 166.66% or 150% (in the case of many exemptions) of the designed pressure, depending on the regulations that apply. For example, if a cylinder was rated to DOT-2015 PSI (approximately 139 bar), it would be tested at around 3360 PSI (approximately 232 bar). Water is commonly used because it is almost incompressible (compressible only by weight, not air pressure), so will only expand by a very small amount should the vessel split. If high pressure gas were used, then the gas would expand to pV=nRT its compressed volume in an explosion, with the attendant risk of damage or injury. This is the risk which the testing is intended to mitigate. In other words using water is safer and takes less energy to do. Just one drop of water in every 5 seconds can cause a pressure change even up to 25 psi, while it would take a huge amount of gas to produce the same pressure change. Water is used mainly because the density of water is very close to 1.001 so using scale 1 g of water in the bowl would be 1 ml of cylinder expansion.

Small pressure vessels are normally tested using a water jacket test. The vessel is visually examined for defects and then placed in a container filled with water, and in which the change in volume of the vessel can be measured by monitoring the water level. A gauge is used to measure the amount of change. The vessel is then pressurised for a specified period, usually 30 or more seconds, and then depressurised again.

The water level in the jacket is then examined. The level will be greater if the vessel being tested has been distorted by the pressure change and did not return to its original volume, or some of the pressurised water inside has leaked out. In both cases, this will normally signify that the vessel has failed the test. If the percentage of permanent expansion is more than 10%, or not up to DOT or customer standards, the cylinder fails, and then goes through a condemning process marking the cylinder as unsafe. This measures the overall leakage of a system instead of locating the leaks and additives can be added to the water to reduce resistivity and increase the sensitivity of the test. The hydrostatic test fluid can also clog small holes (1x10^-6 std cm^3/s or smaller) as a result of the increase in pressure. This is another reason why water is commonly used.

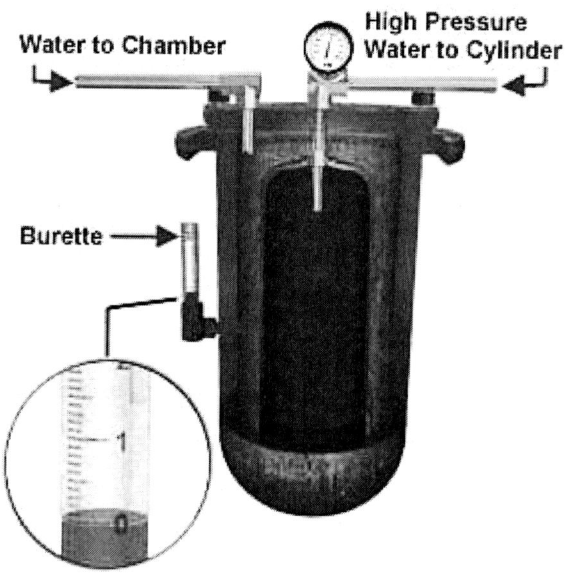

Figure: Water jacket test

All the information the tester needs is stamped onto the cylinder. This includes the DOT information, serial number, manufacturer, and manufacture date. Other information is stamped as needed such as the REE or how much the manufacturer specifies the cylinder should expand before it is considered unsafe. All this information is usually taken down and stored on a computer prior to the testing process. All this information is necessary for keeping track of when the cylinder has been or needs to be hydrotested.

A simpler test, that is still considered a hydrostatic test but can be performed by anyone who has a garden hose, is to pressurise the vessel by filling it with water and to physically examine the outside for leaks. The pressure level achieved in this sort of test does not come close to the pressure level that would be used in a professional testing facility.

Examples

Portable fire extinguishers are safety tools that are required to be on hand in almost every public building. Fire extinguishers are also highly recommended in every home. Over time the conditions in which they are housed, and the manner in which they are handled have an impact on the structural integrity of the extinguisher. A structurally weakened fire extinguisher can malfunction or even burst when it is needed the most. To maintain the quality and safety of this product, hydrostatic testing must be utilised. All critical components of the fire extinguisher must be tested to ensure proper function. The cylinder would be tested by using the water jacket test.

As previously mentioned, the water pressure inside the tank will usually be 150% of the normal operating pressure. The change in volume of the cylinder is calculated by measuring the change in the water levels outside the cylinder. This can be done with a digital scale as well to detect the slightest changes, almost always in grams. The cylinder can also be visually checked for leaks or the pressure drop method can be utilised to measure the overall efficiency of the cylinder.

Pipeline Testing

Hydrotesting of pipe, pipelines and vessels is performed to expose defective materials that have missed prior detection, ensure that any remaining defects are insignificant enough to allow operation at design pressures, expose possible leaks and serve as a final validation of the integrity of the constructed system. ASME B31.3 requires this testing to ensure tightness and strength. ASME B31.3 section 345

Buried high pressure oil and gas pipelines are tested for strength by pressurising them to at least 125% of their maximum operating pressure (MAOP) at any point along their length. Since many long distance transmission pipelines are designed to have a steel hoop stress of 80% of specified minimum yield (SMYS) at MAOP, this means that the steel is stressed to SMYS and above during the testing, and test sections must be selected to ensure that excessive plastic deformation does not occur.

Test pressures need not exceed a value that would produce a stress higher than yield stress at test temperature. ASME B31.3 section 345.4.2 (c)

Other codes require a more onerous approach. BS PD 8010-2 requires testing to 150% of the design pressure-which should not be less than the MAOP plus surge and other incidental effects that will occur during normal operation.

Leak testing is performed by balancing changes in the measured pressure in the test section against the theoretical pressure changes calculated from changes in the measured temperature of the test section.

Australian standard AS2885.5 "Pipelines—Gas and liquid petroleum: Part 5: Field pressure testing" gives an excellent explanation of the factors involved.

Testing Frequency

Most countries have legislation or building code that requires pressure vessels to be regularly tested, for example every two years (with a visual inspection annually) for high pressure gas cylinders and every five or ten years for lower pressure ones such as used in fire extinguishers. Gas cylinders which fail are normally destroyed as part of the testing protocol to avoid the dangers inherent in them being subsequently used.

These common gas cylinders have the following requirements:

- DOT-3AL gas cylinders must be tested every 5 years and have an unlimited life.
- DOT-3HT gas cylinders must be tested every 3 years and have a 24 year life.
- SP-Cylinders Special Permit
- DOT-3AA gas cylinders must be tested every 5 years and have an unlimited life. (Unless stamped with a star (*) in which case the cylinder meets certain specifications and can have a 10 year hydrostatic test life)

In the U S and Canada, organisations such as ASTM, and ASME specify the guidelines for the different types of pressure vessels.

Insulated Pipe

District heating pipline from Lower Austria

Figure: Elbow with foam pads of a District heating pipline with an steel tube diametre of 400 mm installed in Lower Austria

Insulated pipes (called also preinsulated pipes or bonded pipe) are widely used for district heating and hot water supply in Europe. They consist of a steel pipe, an insulating layer, and an outer casing. The main purpose of such pipes is to maintain the temperature of the fluid in the pipes. A common application is the hot water from district heating plants. Most commonly used are single insulated pipes, but more recently in Europe it is becoming popular to use two pipes insulated within the same casing. By using insulated pipe supports, direct heat transfer between pipes and their supports are prevented.

The insulating material usually used is polyurethane foam or similar, with a coefficient of thermal conductivity k=0.033-0.024 W/mK (thermal conductivity). Outer casing is usually high-density polyethylene (HDPE). Production of preinsulated pipes for district heating in the European Union is regulated by the standard EN253. According to EN253:2003, pipes must be produced to work at constant temperature of 130 °C for 30 years, keeping heat conductivity less than or equal to λ=0,033 W/mK. There are three insulation thickness levels.

Insulated pipelines are usually assembled from pipes of 6 m, 12 m, or 16 m in length, laid underground in depth 0.4–1.0 m. Efficient working life of district heating pipelines networks is estimated at 25–30 years, after which they need to be replaced with new pipes.

Four major producers of preinsulated pipes in Europe are located in Denmark and Sweden.

Iron Pipe Size

Iron Pipe Size (IPS) refers to an old pipe sizing system still in use by some industries, including major PVC pipe manufacturers, as well as some legacy drawings and equipment.

The iron pipe size standard came into being early in the 19th century and remained in effect until after World War II. The IPS system was primarily used in the US and the United Kingdom. In the 1920s, the Copper Tube Size (CTS) standard was combined with the IPS standard.

During the IPS period, pipes were cast in halves and welded together, and pipes' dimensions were sized by reference to the outside diametres ("ODs") of the pipes. The ODs under IPS were roughly as we know them today under the Ductile Iron Pipe Standard (DIPS) and Nominal Pipe Size (NPS) Standards, and some of the wall thicknesses were also retained with a different designator. In 1948, the DIPS came into effect, when greater control of a pipe's wall thickness was possible.

CTS diametre always specifies the OD of a tube, where pipe diametre specifications only approximate the pipe ID for sizes of 12 inch or less, and STD wall thickness. The IPS number (reference to an OD) is the same as the NPS number, but the schedules were limited to Standard Wall (STD), Extra Strong, (XS) and Double Extra Strong (XXS). STD is identical to SCH 40 for NPS 1/8 to NPS 10, inclusive, and indicates.375" wall thickness for NPS 12 and larger. XS is identical to SCH 80 for NPS 1/8 to NPS 8, inclusive, and indicates.500" wall thickness for NPS 8 and larger. Different definitions exist for XXS, but it is generally thicker than schedule 160.

Moody Chart

The Moody chart or Moody diagram is a graph in non-dimensional form that relates the Darcy friction factor, Reynolds number and relative roughness for fully developed flow in a circular pipe. It can be used for working out pressure drop or flow rate down such a pipe.

Description

This dimensionless chart is used to work out pressure drop, ΔP (Pa) (or head loss, h_f (m)) and flow rate through pipes.

Head loss can be calculated using the Darcy–Weisbach equation:

$$h_f = f\frac{l}{d}\frac{V^2}{2g};$$

not to be confused with the Fanning equation:

$$h_f = 4f\frac{l}{d}\frac{V^2}{2g},$$

which uses a friction-factor equal to one fourth the Darcy-Weisbach friction factor. Pressure drop can then be evaluated as:

$$\Delta P = \rho g h_f \text{ or directly from } \Delta P = f\frac{\rho V^2}{2}\frac{l}{d},$$

where p is the density of the fluid, V is the average velocity in the pipe, f is the friction factor from the Moody chart, l is the length of the pipe and d is the pipe diametre.

The basic chart plots Darcy–Weisbach friction factor against Reynolds number for a variety of relative roughnesses and flow regimes.

The relative roughness being the ratio of the mean height of roughness of the pipe to the pipe diametre or $\frac{\epsilon}{d}$.

The Moody chart can be divided into two regimes of flow: laminar and turbulent. For the laminar flow regime, the Darcy–Weisbach friction factor was determined analytically by Poiseuille and $\frac{64}{Re}$ is used. In this regime roughness has no discernible effect. For the turbulent flow regime, the relationship between the friction factor and the Reynolds number is more complex and is governed by the Colebrook equation which is implicit in f:

$$\frac{1}{\sqrt{f}} = -2.0\log_{10}\left(\frac{\frac{\epsilon}{d}}{3.7} + \frac{2.51}{Re\sqrt{f}}\right), \text{turbulent flow.}$$

In 1944, Lewis Ferry Moody plotted the Darcy–Weisbach friction factor into what is now known as the Moody chart. The Fanning friction factor is 1/4 the Darcy–Weisbach one and the equation for pressure drop has a compensating factor of four.

Orifice Plate

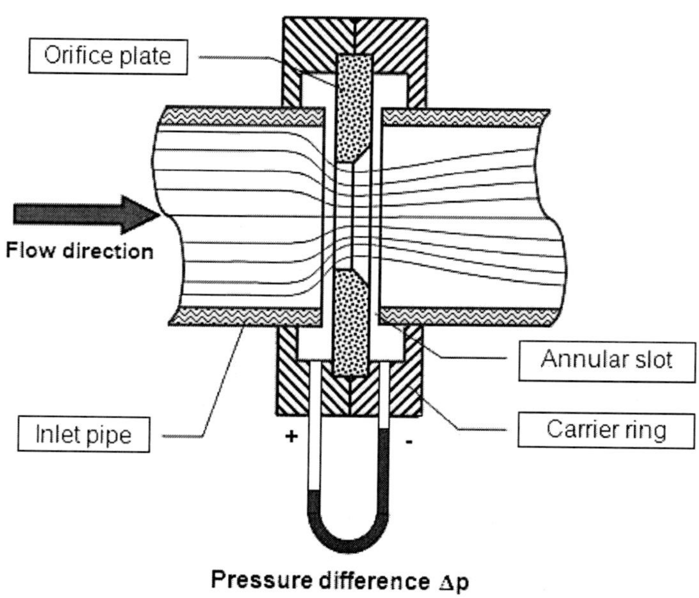

Figure: ISO 5167 Orifice Plate

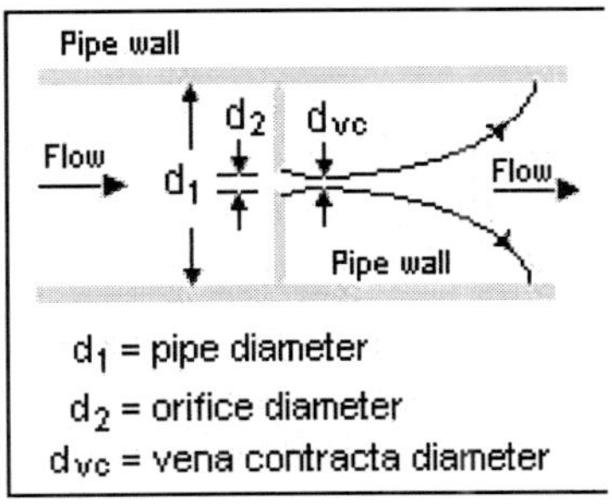

Figure: Flat-plate, sharp-edge orifice

An orifice plate is a device used for measuring the volumetric flow rate. It uses the same principle as a Venturi nozzle, namely Bernoulli's principle which states that there is a relationship between the pressure of the fluid and the velocity of the fluid. When the velocity increases, the pressure decreases and vice versa.

Description

An orifice plate is a thin plate with a hole in the middle. It is usually placed in a pipe in which fluid flows. When the fluid reaches the orifice plate, with the hole in the middle, the fluid is forced to converge to go through the small hole; the point of maximum convergence actually occurs shortly downstream of the physical orifice, at the so-called vena contracta point. As it does so, the velocity and the pressure changes. Beyond the vena contracta, the fluid expands and the velocity and pressure change once again. By measuring the difference in fluid pressure between the normal pipe section and at the vena contracta, the volumetric and mass flow rates can be obtained from Bernoulli's equation.

Uses

Orifice plates are most commonly used for continuous measurement of fluid flow in pipes. They are also used in some small river systems to measure flow rates at locations where the river passes through a culvert or drain. Only a small number of rivers are appropriate for the use of the technology since the plate must remain completely immersed i.e the approach pipe must be full, and the river must be substantially free of debris.

A restrictive flow orifice, a type of orifice plate, is a safety device to control maximum flow from a compressed gas cylinder.

In the natural environment large orifice plates are used to control onward flow in flood relief dams. In these structures a low dam is placed across a river and in normal operation the water flows through the orifice plate unimpeded as the orifice is substantially larger than the normal flow cross section. However, in floods, the flow rate rises and floods out the orifice plate which can then only pass a flow determined by the physical dimensions of the orifice. Flow is then held back behind the low dam in a temporary reservoir which is slowly discharged through the orifice when the flood subsides.

Incompressible Flow Through an Orifice

By assuming steady-state, incompressible (constant fluid density), inviscid, laminar flow in a horizontal pipe (no change in elevation) with negligible frictional losses, Bernoulli's equation reduces to an equation relating the conservation of energy between two points on the same streamline:

$$P_1 + \frac{1}{2}\cdot\rho\cdot V_1^2 = P_2 + \frac{1}{2}\cdot\rho\cdot V_2^2$$

or:

$$P_1 - P_2 = \frac{1}{2}\cdot\rho\cdot V_2^2 - \frac{1}{2}\cdot\rho\cdot V_1^2$$

By continuity equation:

$Q = A_1\cdot V_1 = A_2\cdot V_2$ or $V_1 = Q/A_1$ and $V_2 = Q/A_2$:

$$P_1 - P_2 = \frac{1}{2}\cdot\rho\cdot\left(\frac{Q}{A_2}\right)^2 - \frac{1}{2}\cdot\rho\cdot\left(\frac{Q}{A_1}\right)^2$$

Solving for Q:

$$Q = A_2\sqrt{\frac{2\,(P_1 - P_2)/\rho}{1-(A_2/A_1)^2}}$$

and:

$$Q = A_2\sqrt{\frac{1}{1-(d_2/d_1)^4}}\sqrt{2\,(P_1 - P_2)/\rho}$$

The above expression for Q gives the theoretical volume flow rate. Introducing the beta factor $\beta = d_2/d_1$ as well as the coefficient of discharge C_d:

$$Q = C_d\,A_2\sqrt{\frac{1}{1-\beta^4}}\sqrt{2\,(P_1 - P_2)/\rho}$$

And finally introducing the metre coefficient C which is defined as $C = \frac{C_d}{\sqrt{1-\beta^4}}$ to obtain the final equation for the volumetric flow of the fluid through the orifice:

$$(1) \qquad Q = C\, A_2 \sqrt{2\,(P_1 - P_2)/\rho}$$

Multiplying by the density of the fluid to obtain the equation for the mass flow rate at any section in the pipe:

$$(2) \qquad \dot{m} = \rho\, Q = C\, A_2 \sqrt{2\,\rho\,(P_1 - P_2)}$$

where:

Q	= volumetric flow rate (at any cross-section), m^3/s
m	= mass flow rate (at any cross-section), kg/s
C_d	= coefficient of discharge, dimensionless
C	= orifice flow coefficient, dimensionless
A_1	= cross-sectional area of the pipe, m^2
A_2	= cross-sectional area of the orifice hole, m^2
d_1	= diametre of the pipe, m
d_2	= diametre of the orifice hole, m
β	= ratio of orifice hole diametre to pipe diametre, dimensionless
V_1	= upstream fluid velocity, m/s
V_2	= fluid velocity through the orifice hole, m/s
P_1	= fluid upstream pressure, Pa with dimensions of $kg/(m \cdot s^2)$
P_2	= fluid downstream pressure, Pa with dimensions of $kg/(m \cdot s^2)$
ρ	= fluid density, kg/m^3

Deriving the above equations used the cross-section of the orifice opening and is not as realistic as using the minimum cross-section at the vena contracta. In addition, frictional losses may not be negligible and viscosity and turbulence effects may be present. For that reason, the coefficient of discharge C_d is introduced. Methods exist for determining the coefficient of discharge as a function of the Reynolds number.

The parameter $\sqrt{1-\beta^4}$ is often referred to as the *velocity of approach factor* and dividing the coefficient of discharge by that parameter (as was done above) produces the flow coefficient C. Methods also exist for determining the flow coefficient as a function of the beta function β and the location of the downstream pressure sensing tap. For rough approximations, the flow coefficient may be assumed to be between 0.60 and 0.75. For a first approximation, a flow coefficient of 0.62 can be used as this approximates to fully developed flow.

An orifice only works well when supplied with a fully developed flow profile. This is achieved by a long upstream length (20 to 40 pipe diametres, depending on Reynolds number) or the use of a flow conditioner. Orifice plates are small and inexpensive but do not recover the pressure drop as well as a venturi nozzle does. If space permits, a venturi metre is more efficient than an orifice plate.

Flow of Gases Through an Orifice

In general, equation (2) is applicable only for incompressible flows. It can be modified by introducing the expansion factor Y to account for the compressibility of gases.

$$(3) \qquad \dot{m} = \rho_1\, Q = C\, Y\, A_2\, \sqrt{2\, \rho_1\, (P_1 - P_2)}$$

Y is 1.0 for incompressible fluids and it can be calculated for compressible gases.

Calculation of Expansion Factor

The expansion factor Y, which allows for the change in the density of an ideal gas as it expands is entropically, is given by:

$$Y = \sqrt{r^{2/k}\left(\frac{k}{k-1}\right)\left(\frac{1-r^{(k-1)/k}}{1-r}\right)\left(\frac{1-\beta^4}{1-\beta^4\, r^{2/k}}\right)}$$

For values of β less than 0.25, β^4 approaches 0 and the last bracketed term in the above equation approaches 1. Thus, for the large majority of orifice plate installations:

$$(4) \qquad Y = \sqrt{r^{2/k}\left(\frac{k}{k-1}\right)\left(\frac{1-r^{(k-1)/k}}{1-r}\right)}$$

where:

Y = Expansion factor, dimensionless

r = P_2 / P_1

k = specific heat ratio (), dimensionless

Substituting equation (4) into the mass flow rate equation (3):

$$\dot{m} = C\, A_2 \sqrt{2\, \rho_1 \left(\frac{k}{k-1}\right)\left[\frac{(P_2/P_1)^{2/k} - (P_2/P_1)^{(k+1)/k}}{1 - P_2/P_1}\right](P_1 - P_2)}$$

and:

$$\dot{m} = C\, A_2 \sqrt{2\, \rho_1 \left(\frac{k}{k-1}\right)\left[\frac{(P_2/P_1)^{2/k} - (P_2/P_1)^{(k+1)/k}}{(P_1 - P_2)/P_1}\right](P_1 - P_2)}$$

and thus, the final equation for the non-choked (i.e., sub-sonic) flow of ideal gases through an orifice for values of β less than 0.25:

(5) $$\dot{m} = C\, A_2 \sqrt{2\, \rho_1\, P_1 \left(\frac{k}{k-1}\right)\left[(P_2/P_1)^{2/k} - (P_2/P_1)^{(k+1)/k}\right]}$$

Using the ideal gas law and the compressibility factor (which corrects for non-ideal gases), a practical equation is obtained for the non-choked flow of real gases through an orifice for values of β less t h a n 0.25:

(6) $$\dot{m} = C\, A_2\, P_1 \sqrt{\frac{2\,M}{Z\,R\,T_1}\left(\frac{k}{k-1}\right)\left[(P_2/P_1)^{2/k} - (P_2/P_1)^{(k+1)/k}\right]}$$

Remembering that $Q_1 = \dfrac{\dot{m}}{\rho_1}$ and $\rho_1 = M\,\dfrac{P_1}{Z\,R\,T_1}$ (ideal gas law and the compressibility factor)

(8) $$Q_1 = C\, A_2 \sqrt{2\,\frac{Z\,R\,T_1}{M}\left(\frac{k}{k-1}\right)\left[(P_2/P_1)^{2/k} - (P_2/P_1)^{(k+1)/k}\right]}$$

where:

---

k = specific heat ratio (c_p / c_v), dimensionless

m = mass flow rate at any section, kg/s

Q_1 = upstream real gas flow rate, m^3/s

C = orifice flow coefficient, dimensionless

A_2 = cross-sectional area of the orifice hole, m^2

ρ_1 = upstream real gas density, kg/m^3

P_1 = upstream gas pressure, Pa with dimensions of $kg/(m \cdot s^2)$

P_2 = downstream pressure, Pa with dimensions of kg/(m·s²)

M = the gas molecular mass, kg/mol (also known as the molecular weight)

R = the Universal Gas Law Constant = 8.3145 J/(mol·K)

T_1 = absolute upstream gas temperature, K

Z = the gas compressibility factor at P_2 and T_1, dimensionless

A detailed explanation of choked and non-choked flow of gases, as well as the equation for the choked flow of gases through restriction orifices, is available at Choked flow.

The flow of real gases through thin-plate orifices never becomes fully choked. "Cunningham (1951) first drew attention to the fact that choked flow will not occur across a standard, thin, square-edged orifice." The mass flow rate through the orifice continues to increase as the downstream pressure is lowered to a perfect vacuum, though the mass flow rate increases slowly as the downstream pressure is reduced below the critical pressure.

Permanent Pressure Drop for Incompressible Fluids

For a square-edge orifice plate with flange taps:

$$\frac{\Delta P_p}{\Delta P_i} = 1 - 0.24\beta - 0.52\beta^2 - 0.16\beta^3$$

where:

ΔP_p = permanent pressure drop

ΔP_i = indicated pressure drop at the flange taps

$\beta = d_2 / d_1$

And rearranging the formula near the top of this article:

$$\Delta P_i = P_1 - P_2 = \frac{Q^2 \rho (1-\beta^4)}{2C_d^2 A_2^2} = \frac{Q^2 \rho (1-\beta^4)}{2C_d^2 A_1^2 \beta^4}$$

Chapter 4

Pipe Bursting

Pipe bursting is a trenchless method of replacing buried pipelines (such as sewer, water, or natural gas pipes) without the need for a traditional construction trench. "Launching and receiving pits" replace the trench needed by conventional pipe-laying.

Methodology

An expanding device called an expander head, which may be either pneumatic or hydraulic, is introduced into the defective pipeline through a launching pit. As it travels through the pipeline toward the receiving pit it breaks the pipe into many small pieces, pushing the pieces into the surrounding soil. New pipe is attached to the back of the expander head, replacing the line immediately.

Equipment

There are five key pieces of equipment used in a pipebursting operation: the expander head, pulling rods, a pulling machine, a retaining device, and a hydraulic power pack. Today's expander heads have a leading end much smaller in diametre than the trailing (bursting) end, small enough to fit through the pipe that will be replaced. The smaller leading end is designed to guide the expander head through the existing pipe; earlier models did not have this feature and lost course at times, resulting in incomplete pipe bursts and project failures. The transition from the leading end to the trailing end can include "fins" that make first contact with the existing pipe. Using these fins as the primary breaking point is a very effective way to ensure that the pipe is broken along the entire circumference. A machine is set in the receiving pit to pull the expander head and new pipe into the line. The head is pulled by heavy, interlocking links that form a chain. Each link weighs several hundred pounds.

All of the equipment used in a pipebursting operation is powered by one or multiple hydraulic power generators.

Other Applications

Pipebursting may also be used to expand pipeline carrying capacity by replacing smaller pipes with larger ones, or "upsizing." Extensive proving work by the gas and water industries has demonstrated the feasibility of upsizing gas mains, water mains and sewers. Upsizing from 100mm to 225mm diametre is now well established, and pipes of up to 900mm diametre and greater have been replaced.

Trenchless Technology

Trenchless technology is a type of subsurface construction work that requires few trenches or no continuous trenches. It is a rapidly growing sector of the construction and civil engineering industry. It can be defined as "a family of methods, materials, and equipment capable of being used for the installation of new or replacement or rehabilitation of existing underground infrastructure with minimal disruption to surface traffic, business, and other activities."

Trenchless Construction

Trenchless construction includes such construction methods as tunnelling, microtunnelling (MTM), horizontal directional drilling (HDD) also known as directional boring, pipe ramming (PR), pipe Jacking (PJ), moling, horizontal auger boring (HAB) and other methods for the installation of pipelines and cables below the ground with minimal excavation. Large diametre tunnels such as those constructed by a tunnel boring machine (TBM), and drilling and blasting techniques are larger versions of subsurface construction. The difference between trenchless and other subsurface construction techniques depends upon the size of the passage under construction.

The method requires considering soil characteristics and the loads applied to the surface. In cases where the soil is sandy, the water table is at shallow depth, or heavy loads like that of urban traffic are expected, the depth of excavation has to be at a depth such that the pressure of the load on the surface does not affect the bore, otherwise there is danger of surface caving in.

Trenchless Rehabilitation

Trenchless rehabilitation includes such construction methods as sliplining, thermoformed pipe, pipe bursting, shotcrete, gunite, cured-in-place pipe (CIPP), grout-in-place pipe, mechanical spot repair, and other methods for the repair, rehabilitation, or replacement of existing buried pipes and structures without excavation, or at least with minimal excavation. Mechanical Spot Repair is applied where damaged pipelines require the re-instatement of structural integrity. Sliplining, CIPP, and thermoformed pipe lining involve pulling or inverting a new liner into an existing pipe, then applying heat and/or pressure to force the liner to expand to fill the pipe. CIPP technologies combine a carrier (felt or fibreglass) impregnated with heat, ultraviolet light, or ambient curable resin to form a "pipe within a pipe". Pipe bursting fractures a pipe from the inside and forces the fragments outwards while a new pipe is drawn in to replace the old. The other methods are primarily for fixing spot leaks. Trenchless rehabilitation methods are generally more cost-effective than traditional exhume (dig) and replace methods.

Trade Associations

The trenchless technology industry is represented by several industry associations in addition to standing committees within almost every water and sewer related industry association. The National Association of Sewer Service Companies (NASSCO) was established in 1976 and is the oldest such association with a trenchless focus. The North American Society for Trenchless Technology (NASTT) was established in 1990. The Indian Society for Trenchless Technology or IndSTT is the apex organisation to promote Trenchless Technology in South Asia. It was established in 1995 with its head office at New Delhi. The International Society for Trenchless Technology, established in 1986, has 25 active national trenchless technology societies associated with it. Several organisations bring out Technical Journals for sensitising the readers about such techniques. No Dig India is the official Journal of IndSTT.

Pipe Cutting

Pipe cutting, or pipe profiling, is a mechanised industrial process that removes material from pipe or tube to create a desired profile. Typical profiles include straight cuts, mitres, saddles and mid-section holes. These complex cuts are usually required to allow a tight fit between two parts that are to be joined via arc welding. Cutting is performed by means of a thermal torch (plasma or oxyfuel) and is mounted to the last axis of a multi-axis machine. The axes of the multi-axis machine are powered by electric motors and are synchronised to create a path for the torch and pipe that yield a desired profile. The synchronisation of axes is accomplished either mechanically, via cams, levers and gears, or electronically, via microprocessors and controllers.

Pipe cutting machines are popular in offshore, pipe processing, ship building, pressure vessel, structural and mechanical contracting manufacturing because of the complex cuts and profiles typical required in their respective industries.

Pipe Drift

Pipe drift is a measure of the roundness or eccentricity of the inside wall of a pipe. "API drift" refers to primary specifications set forth in API Specification 5CT (ISO 11960), "Specification for Casing and Tubing". "Alternate drift" refers to alternate drift specifications listed in API 5CT. "Special drift" refers to industry drift specifications other than those found in API 5CT.

Drifting means measuring a pipe's inner roundness. Drifting is typically performed by passing a cylindrical mandrel through the length of the pipe to detect occlusions. It occurs both at the pipe mill and in the field. Drifting is performed to ensure that tools, pumps, smaller pipe, and other items can be passed through the pipe.

Pipe Flow

Pipe flow, a branch of Hydraulics and Fluid Mechanics, is a type of liquid flow within a closed conduit (conduit in the sense of a means of containment). The other type of flow within a conduit is open channel flow. These two types of flow are similar in many ways, but differ in one important respect. Pipe flow does not have a free surface which is found in open-channel flow. Pipe flow, being confined within closed conduit, does not exert direct atmospheric pressure, but does exert hydraulic pressure on the conduit.

Not all flow within a closed conduit is considered pipe flow. Storm sewers are closed conduits and usually maintain a free surface and therefore are considered open-channel flow. The exception to this is when a storm sewer operates at full capacity, and then can become pipe flow. Energy in pipe flow is expressed as head and is defined by the Bernoulli equation. In order to conceptualise head along the course of flow within a pipe, diagrams often contain a hydraulic grade line. Pipe flow is subject to frictional losses as defined by the Darcy-Weisbach formula.

States of Flow

The beahviour of pipe flow is governed mainly by the effects of viscosity and gravity relative to the inertial forces of the flow. Depending on the effect of viscosity relative to inertia, as represented by the Reynolds number, the flow can be either laminar or turbulent. At a

Reynolds number below the critical value of approximately 2040 pipe flow will ultimately be laminar, whereas above the critical value turbulent flow can persist. In addition, the transition between laminar flow and turbulence can be sensitive to disturbance levels and imperfections.

Flow through pipes can roughly be divided into two:

- Laminar flow
- Turbulent flow.

Hagen–Poiseuille Equation

In fluid dynamics, the Hagen–Poiseuille equation is a physical law that gives the pressure drop in a fluid flowing through a long cylindrical pipe. The assumptions of the equation are that the flow is laminar viscous and incompressible and the flow is through a constant circular cross-section that is substantially longer than its diametre. The equation is also known as the Hagen–Poiseuille law, Poiseuille law and Poiseuille equation.

The fluid flow will be turbulent for velocities and pipe diametres above a threshold, leading to larger pressure drops than would be expected according to the Hagen–Poiseuille equation.

Equation

Standard Fluid Dynamics Notation

In standard fluid dynamics notation:

$$\Delta P = \frac{8\mu LQ}{\pi r^4}$$

or

$$\Delta P = \frac{128\mu LQ}{\pi d^4}$$

where:

ΔP is the pressure drop

$L$is the length of pipe

μ is the dynamic viscosity

Q is the volumetric flow rate

r is the radius

d is the diametre

π is the mathematical constant.

Physics Notation

$$\Phi = \frac{dV}{dt} = v\pi R^2 = \frac{\pi R^4}{8\eta}\left(\frac{-\Delta P}{\Delta x}\right) = \frac{\pi R^4}{8\eta}\frac{|\Delta P|}{L}$$

where:

$\phi$is the volumetric flow rate (denoted as Q above)

V is a volume of the liquid poured (cubic metres)

t is the time (seconds)

u is mean fluid velocity along the length of the tube (metres/second)

x is a distance in direction of flow (metres)

R is the internal radius of the tube (metres)

ΔP is the pressure difference between the two ends (pascals)

η is the dynamic fluid viscosity (pascal-second (Pa·s)),

$L$is the total length of the tube in the x direction (metres).

The equation does not hold close to the pipe entrance.

Relation to Darcy–Weisbach

This result is also a solution to the phenomenological Darcy–Weisbach equation in the field of hydraulics, given a relationship for the friction factor in terms of the Reynolds number:

$$\Lambda = \frac{64}{\mathrm{Re}}, \qquad \mathrm{Re} = \frac{2\rho v r}{\eta},$$

where Re is the Reynolds number and ρ fluid density. In this form the law approximates the *Darcy friction factor*, the *energy (head) loss factor*, *friction loss factor* or *Darcy (friction) factor* $\wedge$ in the laminar flow at very low velocities in cylindrical tube. The theoretical derivation of a slightly different form of the law was made independently by Wiedman in 1856 and Neumann and E. Hagenbach in 1858 (1859, 1860). Hagenbach was the first who called this law the Poiseuille's law. The law is also very important specially in hemorheology and hemodynamics, both fields of physiology.

The Poiseuilles' law was later in 1891 extended to turbulent flow by L. R. Wilberforce, based on Hagenbach's work.

Derivation

The Hagen–Poiseuille equation can be derived from the Navier–Stokes equations.

Viscosity

Figure: Two fluids moving past each other in the x direction. The liquid on top is moving faster and will be pulled in the negative direction by the bottom liquid while the bottom liquid will be pulled in the positive direction by the top liquid.

The derivation of Poiseuille's law is surprisingly simple, but it requires an understanding of viscosity. When two layers of liquid in contact with each other move at different speeds, there will be a shear force between them. This force is proportional to the area of contact A, the velocity gradient in the direction of flow $\Delta v_x / \Delta y$, and a proportionality constant η (viscosity) and is given by

$$F_{\text{viscosity, top}} = -\eta A \frac{\Delta v_x}{\Delta y}.$$

The negative sign is in there because we are concerned with the faster moving liquid, which is being slowed by the slower liquid. By Newton's third law of motion, the force on the slower liquid is equal and opposite (no negative sign) to the force on the faster liquid. This equation assumes that the area of contact is so large that we can ignore any effects from the edges and that the fluids behave as Newtonian fluids.

Liquid Flow Through a Pipe

In a tube we make a basic assumption: the liquid in the centre is moving fastest while the liquid touching the walls of the tube is stationary (due to friction). To simplify the situation, let's assume that there are a bunch of circular layers (lamina) of liquid, each having a velocity determined only by their radial distance from the centre of the tube.

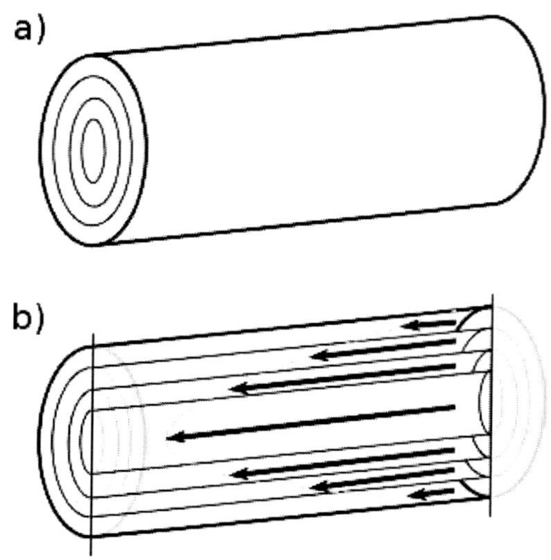

Figure: a) A tube showing the imaginary lamina. b) A cross section of the tube shows the lamina moving at different speeds. Those closest to the edge of the tube are moving slowly while those near the centre are moving quickly.

To figure out the motion of the liquid, we need to know all forces acting on each lamina:

1. The force pushing the liquid through the tube is the change in pressure multiplied by the area: $F = -\Delta PA$. This force is in the direction of the motion of the liquid-the negative sign comes from the conventional way we define.

$$\Delta P = P_{end} - P_{top} < 0$$

2. The pull from the faster lamina immediately closer to the centre of the tube
3. The drag from the slower lamina immediately closer to the walls of the tube.

The first of these forces comes from the definition of pressure. The other two forces require us to modify the equations above that we have for viscosity. In fact, we are not modifying the equations, instead merely plugging in values specific to our problem. Let's focus on the pull from the faster lamina (#2) first.

Faster Lamina

From the equation above, we need to know the area of contact and the velocity gradient. Think of the lamina as a ring of radius and thickness. The area of contact between the lamina and the faster one is simply the area of the inside of the cylinder: $A = 2\pi s\Delta x$. We don't know the exact form for the velocity of the liquid within the tube yet, but we do know (from our assumption above) that it is dependent on the radius. Therefore, the velocity gradient is the change of the velocity with respect to the change in the radius at the intersection of these two laminae. That intersection is at a radius of. So, considering that this force will be positive with respect to the movement of the liquid (but the derivative of the velocity is negative), the final form of the equation becomes

$$F_{\text{viscosity, fast}} = -\eta 2\pi s \Delta x \left.\frac{dv}{dr}\right|_s$$

where the vertical bar and subscript s following the derivative indicates that it should be taken at a radius of.

Slower Lamina

Next let's find the force of drag from the slower lamina. We need to calculate the same values that we did for the force from the faster lamina. In this case, the area of contact is at $s+ds$ instead of s. Also, we need to remember that this force opposes the direction of movement of the liquid and will therefore be negative (and that the derivative of the velocity is negative).

$$F_{\text{viscosity, slow}} = \eta 2\pi (s+ds)\Delta x \left.\frac{dv}{dr}\right|_{s+ds}$$

Putting it All Together

To find the solution for the flow of liquid through a tube, we need to make one last assumption. There is no acceleration of liquid in the pipe, and by Newton's first law, there is no net force.

If there is no net force then we can add all of the forces together to get zero

$$0 = F_{\text{pressure}} + F_{\text{viscosity, fast}} + F_{\text{viscosity, slow}}$$

or

$$0 = -\Delta P 2\pi s ds - \eta 2\pi s \Delta x \left.\frac{dv}{dr}\right|_s + \eta 2\pi (s+ds)\Delta x \left.\frac{dv}{dr}\right|_{s+ds}.$$

First, to get everything happening at the same point, use the first two terms of a Taylor series expansion of the velocity gradient:

$$\left.\frac{dv}{dr}\right|_{r+dr} = \left.\frac{dv}{dr}\right|_r + \left.\frac{d^2v}{dr^2}\right|_r dr.$$

Also, use r instead of s since the lamina was arbitrary and the expression must be valid for all laminae. Grouping like terms and dropping the vertical bar since all derivatives are assumed to be at radius r,

$$0 = -\Delta P 2\pi r dr + \eta 2\pi dr \Delta x \frac{dv}{dr} + \eta 2\pi r dr \Delta x \frac{d^2v}{dr^2} + \eta 2\pi (dr)^2 \Delta x \frac{d^2v}{dr^2}.$$

Finally, put this expression in the form of a differential equation, dropping the term quadratic in dr.

$$\frac{1}{\eta}\frac{\Delta P}{\Delta x} = \frac{d^2v}{dr^2} + \frac{1}{r}\frac{dv}{dr}$$

It can be seen that both sides of the equations are negative: there is a drop of pressure along the tube (left side) and both first and second derivatives of the velocity are negative (velocity has a maximum value of the centre of the tube). Using the chain rule, the equation may be re-arranged to:

$$\frac{1}{\eta}\frac{\Delta P}{\Delta x} = \frac{1}{r}\frac{d}{dr} r \frac{dv}{dr}.$$

This differential equation is subject to the following boundary conditions:

$v(r) = 0$ at $r = R$— "no-slip" boundary condition at the wall

$\frac{dv}{dr} = 0$ at $r = 0$— axial symmetry.

Axial symmetry means that the velocity v(r) is maximum at the centre of the tube, therefore the first derivative $\frac{dv}{dr}$ is zero at r = 0.

The differential equation can be integrated to:

$$v(r) = \frac{1}{4\eta} r^2 \frac{\Delta P}{\Delta x} + A\ln(r) + B.$$

To find A and B, we use the boundary conditions.

First, the symmetry boundary condition indicates:

$$\frac{dv}{dr} = \frac{1}{2\eta} r \frac{\Delta P}{\Delta x} + A\frac{1}{r} = 0 \text{ at } r = 0.$$

A solution possible only if A = 0. Next the no-slip boundary condition is applied to the remaining equation:

$$v(R) = \frac{1}{4\eta} R^2 \frac{\Delta P}{\Delta x} + B = 0$$

so therefore

$$B = -\frac{1}{4\eta} R^2 \frac{\Delta P}{\Delta x}.$$

Now we have a formula for the velocity of liquid moving through the tube as a function of the distance from the centre of the tube

$$v = -\frac{1}{4\eta} \frac{\Delta P}{\Delta x} (R^2 - r^2)$$

or, at the centre of the tube where the liquid is moving fastest ($r = 0$) with R being the radius of the tube,

$$v_{max} = -\frac{1}{4\eta} \frac{\Delta P}{\Delta x} R^2.$$

Poiseuille's Law

To get the total volume that flows through the tube, we need to add up the contributions from each lamina.

To calculate the flow through each lamina, we multiply the velocity (from above) and the area of the lamina.

$$\Phi(r)dr = \frac{1}{4\eta} \frac{|\Delta P|}{\Delta x} (R^2 - r^2) 2\pi r dr = \frac{\pi}{2\eta} \frac{|\Delta P|}{\Delta x} (rR^2 - r^3) dr$$

Finally, we integrate over all lamina via the radius variable r.

$$\Phi = \frac{\pi}{2\eta} \frac{|\Delta P|}{\Delta x} \int_0^R (rR^2 - r^3) dr = \frac{|\Delta P| \pi R^4}{8\eta \Delta x}$$

Poiseuille's Equation for Compressible Fluids

For a compressible fluid in a tube the volumetric flow rate and the linear velocity is not constant along the tube. The flow is usually expressed at outlet pressure.

As fluid is compressed or expands, work is done and the fluid is heated and cooled. This means that the flow rate depends on the heat transfer to and from the fluid.

For an ideal gas in the isothermal case, where the temperature of the fluid is permitted to equilibrate with its surroundings, and when the pressure difference between ends of the pipe is small, the volumetric flow rate at the pipe outlet is given by

$$\Phi = \frac{dV}{dt} = v\pi R^2 = \frac{\pi R^4 \left(P_i - P_o\right)}{8\eta L} \times \frac{P_i + P_o}{2P_o} = \frac{\pi R^4}{16\eta L}\left(\frac{P_i^2 - P_o^2}{P_o}\right)$$

where:

P_i inlet pressure

P_o outlet pressure

$L$is the length of tube

$\eta$is the viscosity

$R$is the radius

V is the volume of the fluid at outlet pressure

u is the velocity of the fluid at outlet pressure

This is usually a good approximation when the flow velocity is less than mach 0.3

This equation can be seen as Poiseuille's law with an extra correction factor $\frac{P_i + P_o}{2} \times \frac{1}{P_o}$ expressing the average pressure relative to the outlet pressure.

Electrical Circuits Analogy

Electricity was originally understood to be a kind of fluid. This hydraulic analogy is still conceptually useful for understanding circuits. This analogy is also used to study the frequency response of fluid mechanical networks using circuit tools, in which case the fluid network is termed a hydraulic circuit.

Poiseuille's law corresponds to Ohm's law for electrical circuits ($V = IR$), where the pressure drop ΔP is analogous to the voltage V and volumetric flow rate ϕ is analogous to the current I. Then the resistance

$$R = \frac{8\eta\Delta x}{\pi r^4}.$$

This concept is useful because the effective resistance in a tube is inversely proportional to the fourth power of the radius. This means that halving the radius of the tube increases the resistance to fluid movement by a factor of 16.

History

It was developed independently by Gotthilf Heinrich Ludwig Hagen (1797-1884) and Jean Louis Marie Poiseuille.

Poiseuille's law was experimentally derived in 1838 and formulated and published in 1840 and 1846 by Jean Louis Marie Poiseuille (1797–1869). Hagen did his experiments in 1839.

Turbulence

n fluid dynamics, turbulence or turbulent flow is a flow regime characterised by chaotic and stochastic property changes. This includes low momentum diffusion, high momentum convection, and rapid variation of pressure and velocity in space and time. Nobel Laureate Richard Feynman described turbulence as "the most important unsolved problem of classical physics." Flow in which the kinetic energy dies out due to the action of fluid molecular viscosity is called laminar flow. While there is no theorem relating Reynolds number (Re) to turbulence, flows at Reynolds numbers larger than 5000 are typically (but not necessarily) turbulent, while those at low Reynolds numbers usually remain laminar. In pipe flow, for example, turbulence can first be sustained if the Reynolds number is larger than a critical value of about 2040; moreover, the turbulence is generally interspersed with laminar flow until a larger Reynolds number of about 3000. In turbulent flow, unsteady vortices appear on many scales and interact with each other. Drag due to boundary layer skin friction increases. The structure and location of boundary layer separation often changes, sometimes resulting in a reduction of overall drag. Although laminar-turbulent transition is not governed by Reynolds number, the same transition occurs if the size of the object is gradually increased, or the viscosity of the fluid is decreased, or if the density of the fluid is increased.

Features

Turbulence is highly characterised by the following features:

Irregularity: Turbulent flows are always highly irregular. This is why turbulence problems are always treated statistically rather than deterministically. Turbulent flow is always chaotic but not all chaotic flows are turbulent.

Diffusivity: Turbulence is highly associated with rapid mixing. One of the useful effects of turbulence, it tends to accelerate the homogenisation of any non-uniform fluid mixture. The process which brings any non-uniform state of a system into a uniform one is called mixing and when the system is in its uniform state, the system becomes a homogeneous system. A mixing process requires sufficient input of energy which is readily available in a turbulent flow. The characteristic which is responsible for the enhanced mixing and increased rates of mass, momentum and energy transports in a flow is regarded as diffusivity.

Turbulent diffusion is usually described by a turbulent diffusion coefficient. This turbulent diffusion coefficient is defined in a phenomenological sense, by analogy with the molecular diffusivities, but it does not have a true physical meaning, being dependent on the flow conditions, and not a property of the fluid itself. In addition, the turbulent diffusivity concept assumes a constitutive relation between a turbulent flux and the gradient of a mean variable similar to the relation between flux and gradient that exists for molecular transport. In the best case, this assumption is only an approximation. Nevertheless, the turbulent diffusivity is the simplest approach for quantitative analysis of turbulent flows, and many models have been postulated to calculate it. For instance, in large bodies of water like oceans this coefficient can be found using Richardson's four-third power law and is governed by the random walk principle. In rivers and large ocean currents, the diffusion coefficient is given by variations of Elder's formula.

Rotationality: Turbulent flows have non-zero vorticity and are characterised by a strong three-dimensional vortex generation mechanism known as vortex stretching. In fluid dynamics, they are essentially vortices subjected to stretching associated with a corresponding increase of the component of vorticity in the stretching direction—due to the conservation of angular momentum. On the other hand, vortex stretching is the core mechanism on which the turbulence energy cascade relies to establish the structure function. In general, the stretching mechanism implies thinning of the vortices in the direction perpendicular to the stretching direction due to volume conservation of fluid elements. As a result, the radial length scale of the vortices decreases and the larger flow structures break down into smaller structures. The process continues until the small scale structures are small enough to the extent where their kinetic energy is overwhelmed by the fluid's molecular viscosity and dissipated into heat. This is why turbulence is always rotational and three dimensional. For example, atmospheric cyclones are rotational but their substantially two-dimensional shapes do not allow vortex generation and so are not turbulent. On the other hand, oceanic flows are dispersive but essentially non rotational and therefore are not turbulent.

Dissipation: To sustain turbulent flow, a constant source of energy supply is required. Otherwise, turbulence dissipates rapidly as the kinetic energy is converted into internal energy by viscous shear stress. Turbulence causes the formation of eddies of many different length scales. Most of the kinetic energy of the turbulent motion is contained in the large-scale structures. The energy "cascades" from these large-scale structures to smaller scale structures by an inertial and essentially inviscid mechanism. This process continues, creating smaller and smaller structures which produces a hierarchy of eddies. Eventually this process creates structures that are small enough that molecular diffusion becomes important and viscous dissipation of energy finally takes place. The scale at which this happens is the Kolmogorov length scale.

Energy cascade: Turbulent flow can be realised as a superposition of a spectrum of velocity fluctuations and eddies on an over mean flow. The eddies are loosely defined as coherent patterns of velocity, vorticity and pressure. Turbulent flows may be viewed as made of an entire hierarchy of eddies over a wide range of length scales and the hierarchy can be described by the energy spectrum that measures the energy in velocity fluctuations for each wave number. The scales in the energy cascade are generally uncontrollable and highly non-symmetric. Nevertheless, based on these length scales these eddies can be divided into three categories.

Integral length scales: Largest scales in the energy spectrum. These eddies obtain energy from the mean flow and also from each other. Thus these are the energy production eddies which contain the most of the energy. They have the large velocity fluctuation and are low in frequency. Integral scales are highly anisotropic and are defined in terms of the normalised two-point velocity correlations. The maximum length of these scales is constrained by the characteristic length of the apparatus. For example, the largest integral length scale of pipe flow is equal to the pipe diametre. In the case of atmospheric turbulence, this length can reach up to the order of several hundreds kilometres.

Kolmogorov length scales: Smallest scales in the spectrum that form the viscous sub-layer range. In this range, the energy input from nonlinear interactions and the energy drain from viscous dissipation are in exact balance. The small scales are in high frequency which is why turbulence is locally isotropic and homogeneous.

Taylor microscales: The intermediate scales between the largest and the smallest scales which make the inertial subrange. Taylor micro-scales are not dissipative scale but passes down the energy from the largest to the smallest without dissipation. Some literatures do not consider Taylor micro-scales as a characteristic length scale and consider the energy cascade contains only the largest and smallest scales; while the later accommodate both the inertial sub-range and the viscous-sub layer. Nevertheless, Taylor micro-scales is often used in describing the term "turbulence" more conveniently as these Taylor micro-scales play a dominant role in energy and momentum transfer in the wavenumber space.

Although it is possible to find some particular solutions of the Navier-Stokes equations governing fluid motion, all such solutions are unstable at large Reynolds numbers. Sensitive dependence on the initial and boundary conditions makes fluid flow irregular both in time and in space so that a statistical description is needed. Russian mathematician Andrey Kolmogorov proposed the first statistical theory of turbulence, based on the aforementioned notion of the energy cascade (an idea originally introduced by Richardson) and the concept of self-similarity. As a result, the Kolmogorov microscales were named after him. It is now known that the self-similarity is broken so the statistical description is presently modified. Still, a complete description of turbulence remains one of the unsolved problems in physics.

According to an apocryphal story, Werner Heisenberg was asked what he would ask God, given the opportunity. His reply was: "When I meet God, I am going to ask him two questions: Why relativity? And why turbulence? I really believe he will have an answer for the first." A similar witticism has been attributed to Horace Lamb (who had published a noted text book on Hydrodynamics)—his choice being quantum electrodynamics (instead of relativity) and turbulence. Lamb was quoted as saying in a speech to the British Association for the Advancement of Science, "I am an old man now, and when I die and go to heaven there are two matters on which I hope for enlightenment. One is quantum electrodynamics, and the other is the turbulent motion of fluids. And about the former I am rather optimistic."

A more detailed presentation of turbulence with emphasis on high-Reynolds number flow, intended for a general readership of physicists and applied mathematicians, is found in the Scholarpedia articles by R. Benzi and U. Frisch. and by G. Falkovich.

Examples of Turbulence

- Smoke rising from a cigarette is turbulent flow. For the first few centimetres, the flow is certainly laminar. Then smoke becomes turbulent as its Reynolds number increases, as its velocity and characteristic length are both increasing.

- Flow over a golf ball. (This can be best understood by considering the golf ball to be stationary, with air flowing over it.) If the golf ball were smooth, the boundary layer flow over the front of the sphere would be laminar at typical conditions. However, the boundary layer would separate early, as the pressure gradient switched from favourable (pressure decreasing in the flow direction) to unfavourable (pressure increasing in the flow direction), creating a large region of low pressure behind the ball that creates high form drag. To prevent this from happening, the surface is dimpled to perturb the boundary layer and promote transition to turbulence. This results in higher skin friction, but moves the point of boundary layer separation further along, resulting in lower form drag and lower overall drag.
- The mixing of warm and cold air in the atmosphere by wind, which causes clear-air turbulence experienced during aeroplane flight, as well as poor astronomical seeing.
- Most of the terrestrial atmospheric circulation.
- The oceanic and atmospheric mixed layers and intense oceanic currents.
- The flow conditions in many industrial equipment (such as pipes, ducts, precipitators, gas scrubbers, dynamic scraped surface heat exchangers, etc.) and machines (for instance, internal combustion engines and gas turbines).
- The external flow over all kind of vehicles such as cars, aeroplanes, ships and submarines.
- The motions of matter in stellar atmospheres.
- A jet exhausting from a nozzle into a quiescent fluid. As the flow emerges into this external fluid, shear layers originating at the lips of the nozzle are created. These layers separate the fast moving jet from the external fluid, and at a certain critical Reynolds number they become unstable and break down to turbulence.

Unsolved Problems in Physics

Is it possible to make a theoretical model to describe the beahviour of a turbulent flow — in particular, its internal structures?

- Race cars unable to follow each other through fast corners due to turbulence created by the leading car causing understeer.

- In windy conditions, trucks that are on the motorway gets buffeted by their wake.
- Bridge supports (piers) in water. In the late summer and fall, when river flow is slow, water flows smoothly around the support legs. In the spring, when the flow is faster, a higher Reynolds Number is associated with the flow. The flow may start off laminar but is quickly separated from the leg and becomes turbulent.
- In many geophysical flows (rivers, atmospheric boundary layer), the flow turbulence is dominated by the coherent structure activities and associated turbulent events. A turbulent event is a series of turbulent fluctuations that contain more energy than the average flow turbulence. The turbulent events are associated with coherent flow structures such as eddies and turbulent bursting, and they play a critical role in terms of sediment scour, accretion and transport in rivers as well as contaminant mixing and dispersion in rivers and estuaries, and in the atmosphere.
- In the medical field of cardiology, a stethoscope is used to detect heart sounds and bruits, which are due to turbulent blood flow. In normal individuals, heart sounds are a product of turbulent flow as heart valves close. However, in some conditions turbulent flow can be audible due to other reasons, some of them pathological. For example, in advanced atherosclerosis, bruits (and therefore turbulent flow) can be heard in some vessels that have been narrowed by the disease process.

Heat and Momentum Transfer

When flow is turbulent, particles exhibit additional transverse motion which enhances the rate of energy and momentum exchange between them thus increasing the heat transfer and the friction coefficient. Assume for a two-dimensional turbulent flow that one was able to locate a specific point in the fluid and measure the actual velocity $v = \left(v_x, v_y\right)$ of every particle that passed through that point at any given time. Then one would find the actual velocity fluctuating about a mean value:

$$v_x = \underbrace{\overline{v_x}}_{\text{mean value}} + \underbrace{v'_x}_{\text{fluctuation}}, \quad v_y = \overline{v_y} + v'_y$$

and similarly for temperature $\left(T = \overline{T} + T'\right)$ and pressure $\left(P = \overline{P} + P'\right)$, where the primed quantities denote fluctuations superposed to the mean. This decomposition of a flow variable into a mean value and a turbulent fluctuation was originally proposed by Osborne Reynolds in 1895, and is considered to be the beginning of the systematic mathematical analysis of turbulent flow, as a sub-field of fluid dynamics. While the mean values are taken as predictable variables determined by dynamics laws, the turbulent fluctuations are regarded as stochastic variables.

The heat flux and momentum transfer (represented by the shear stress) in the direction normal to the flow for a given time are

$$q = \underbrace{v'_y \rho c_P T'}_{\text{experimental value}} = -k_{\text{turb}} \frac{\partial \overline{T}}{\partial y}$$

$$\tau = \underbrace{-\rho \overline{v'_y v'_x}}_{\text{experimental value}} = \mu_{\text{turb}} \frac{\partial \overline{v_x}}{\partial y}$$

where is the heat capacity at constant pressure, ρ is the density of the fluid, μ_{turb} is the coefficient of turbulent viscosity and k_{turb} is the turbulent thermal conductivity.

Kolmogorov's Theory of 1941

Richardson's notion of turbulence was that a turbulent flow is composed by "eddies" of different sizes. The sizes define a characteristic length scale for the eddies, which are also characterised by velocity scales and time scales (turnover time) dependent on the length scale. The large eddies are unstable and eventually break up originating smaller eddies, and the kinetic energy of the initial large eddy is divided into the smaller eddies that stemmed from it. These smaller eddies undergo the same process, giving rise to even smaller eddies which inherit the energy of their predecessor eddy, and so on. In this way, the energy is passed down from the large scales of the motion to smaller scales until reaching a sufficiently small length scale such that the viscosity of the fluid can effectively dissipate the kinetic energy into internal energy.

In his original theory of 1941, Kolmogorov postulated that for very high Reynolds number, the small scale turbulent motions are statistically isotropic (i.e. no preferential spatial direction could be discerned). In general, the large scales of a flow are not isotropic, since they are determined by the particular geometrical features of the boundaries (the size characterising the large scales will be denoted as L). Kolmogorov's idea was that in the Richardson's energy cascade this geometrical and directional information is lost, while the scale is reduced, so that the statistics of the small scales has a universal character: they are the same for all turbulent flows when the Reynolds number is sufficiently high.

Thus, Kolmogorov introduced a second hypothesis: for very high Reynolds numbers the statistics of small scales are universally and uniquely determined by the viscosity (ε) and the rate of energy dissipation (). With only these two parameters, the unique length that can be formed by dimensional analysis is

$$\eta = \left(\frac{\nu^3}{\varepsilon} \right)^{1/4}$$

This is today known as the Kolmogorov length scale.

A turbulent flow is characterised by a hierarchy of scales through which the energy cascade takes place. Dissipation of kinetic energy takes place at scales of the order of Kolmogorov length η, while the input of energy into the cascade comes from the decay of the large scales, of order *L*. These two scales at the extremes of the cascade can differ by several orders of magnitude at high Reynolds numbers. In between there is a range of scales (each one with its own characteristic length *r*) that has formed at the expense of the energy of the large ones. These scales are very large compared with the Kolmogorov length, but still very small compared with the large scale of the flow (i.e. $\eta << r << L$). Since eddies in this range are much larger than the dissipative eddies that exist at Kolmogorov scales, kinetic energy is essentially not dissipated in this range, and it is merely transferred to smaller scales until viscous effects become important as the order of the Kolmogorov scale is approached. Within this range inertial effects are still much larger than viscous effects, and it is possible to assume that viscosity does not play a role in their internal dynamics (for this reason this range is called "inertial range").

Hence, a third hypothesis of Kolmogorov was that at very high Reynolds number the statistics of scales in the range $\eta << r << L$ are universally and uniquely determined by the scale *r* and the rate of energy dissipation.

The way in which the kinetic energy is distributed over the multiplicity of scales is a fundamental characterisation of a turbulent flow. For homogeneous turbulence (i.e., statistically invariant under translations of the reference frame) this is usually done by means of the *energy spectrum function* $E(k)$, where k is the modulus of the wavevector corresponding to some harmonics in a Fourier representation of the flow velocity field u(x):

$$\mathbf{u}(\mathbf{x}) = \iiint_{R^3} u(\mathbf{k})e^{i\mathbf{k}\cdot\mathbf{x}}\mathrm{d}^3\mathbf{k},$$

where û(k) is the Fourier transform of the velocity field. Thus, $E(k)\mathrm{d}k$ represents the contribution to the kinetic energy from all the Fourier modes with $k < |\mathrm{k}| < k + \mathrm{d}k$, and therefore,

$$\frac{1}{2}\langle u_i u_i\rangle = \int_0^\infty E(k)\mathrm{d}k,$$

where $1/2\langle u_i u_i\rangle$ is the mean turbulent kinetic energy of the flow. The wavenumber k corresponding to length scale r is $k = 2\pi / r$. Therefore, by dimensional analysis, the only possible form for the energy spectrum function according with the third Kolmogorov's hypothesis is

$$E(k) = C\varepsilon^{2/3}k^{-5/3},$$

where C would be a universal constant. This is one of the most famous results of Kolmogorov 1941 theory, and considerable experimental evidence has accumulated that supports it.

In spite of this success, Kolmogorov theory is at present under revision. This theory implicitly assumes that the turbulence is statistically self-similar at different scales. This essentially means that the statistics are scale-invariant in the inertial range. A usual way of studying turbulent velocity fields is by means of velocity increments:

$$\delta\mathbf{u}(r) = \mathbf{u}(\mathbf{x}+\mathbf{r}) - \mathbf{u}(\mathbf{x});$$

that is, the difference in velocity between points separated by a vector r (since the turbulence is assumed isotropic, the velocity increment depends only on the modulus of r). Velocity increments are useful because they emphasize the effects of scales of the order of the separation r when statistics are computed. The statistical scale-invariance implies that the scaling of velocity increments should occur with a unique scaling exponent β, so that when r is scaled by a factor λ,

$$\delta\mathbf{u}(\lambda r)$$

should have the same statistical distribution as

$$\lambda^{\beta}\delta\mathbf{u}(r),$$

with β independent of the scale r. From this fact, and other results of Kolmogorov 1941 theory, it follows that the statistical moments of the velocity increments (known as *structure functions* in turbulence) should scale as

$$\langle[\delta\mathbf{u}(r)]^{n}\rangle = C_{n}\varepsilon^{n/3}r^{n/3},$$

where the brackets denote the statistical average, and the C_n would be universal constants.

There is considerable evidence that turbulent flows deviate from this beahviour. The scaling exponents deviate from the $n/3$ value predicted by the theory, becoming a non-linear function of the order n of the structure function. The universality of the constants have also been questioned. For low orders the discrepancy with the Kolmogorov $n/3$ value is very small, which explain the success of Kolmogorov theory in regards to low order statistical moments. In particular, it can be shown that when the energy spectrum follows a power law

$$E(k) \propto k^{-p},$$

with $1 < p < 3$, the second order structure function has also a power law, with the form

$$\langle[\delta\mathbf{u}(r)]^{2}\rangle \propto r^{p-1}.$$

Since the experimental values obtained for the second order structure function only deviate slightly from the 2/3 value predicted by Kolmogorov theory, the value for p is very near to 5/3 (differences are about 2%). Thus the "Kolmogorov -5/3 spectrum" is generally observed in turbulence. However, for high order structure functions the difference with the Kolmogorov scaling is significant, and the breakdown of the statistical self-similarity is clear. This beahviour, and the lack of universality of the constants, are related with the phenomenon of intermittency in turbulence. This is an important area of research in this field, and a major goal of the modern theory of turbulence is to understand what is really universal in the inertial range.

Pipe Wrench

The pipe wrench is an adjustable wrench used for turning soft iron pipes and fittings with a rounded surface. The design of the adjustable jaw allows it to rock in the frame, such that any forward pressure on the handle tends to pull the jaws tighter together. Teeth angled in the direction of turn dig into the soft pipe. They are not intended for use on hardened steel hex nuts or other fittings because they would ruin the head; however, if a hex nut is soft enough that it becomes rounded beyond use with standard wrenches, a pipe wrench is sometimes used to break the bolt or nut free. Pipe wrenches are usually sold in the following sizes (by length of handle): 10, 14, 18, 24, 36, and 48 inches, although smaller and larger sizes are available as well. They are usually made of either steel or aluminium. Teeth, and jaw kits (which also contain adjustment rings and springs) can be bought to repair broken wrenches, as this is cheaper than buying a new wrench.

Figure: A modern pipe wrench.

History

The first pipe or stillson wrench was created by Daniel C. Stillson while working as a mechanic at the Walworth Company. On October 12, 1869, U.S. patent #95,744 was issued to Stillson.

In Prince Edward Island, Canada there is some controversy over the true creator of the wrench. In the island folklore, the wrench was developed by Owen “Iney” McCluskey (or McCloskey), although no proof has ever been presented that would suggest that McCluskey and Stillson ever met. On 17 August 1888 the Swedish inventor Johan Petter Johansson took his first patent on the adjustable pipe wrench. The *Swedish Patent Office* issued the patent (SE 5636) again in 1894. The idea emerged after he established his company Enköpings

Mekaniska Verkstad. Back then there was no standard on nut dimensions so each time a trades man was out on a job, he needed a trolley to take a whole set of fixed pipe wrenches with him. Johan Petter Johansson invented a tool that could grip different nuts dimensions. As a consequence, the adjustable pipe wrench was developed.

Pipe Wrench in Different Languages

In countries such as Australia and the United Kingdom, the previously common term "Stillsons" or "set of Stillsons" is being progressively replaced in everyday speech by the colloquial term monkey wrenches.

This usage is not technically correct, as the term "monkey wrench" refers to a now uncommon type of adjustable wrench that was designed for use with nuts rather than with round pipes.

In South Africa, the terms "bobbejaan spanner" and "baboon spanner" are commonly used, especially for large pipe wrenches. "Bobbejaan" is the Afrikaans term for a baboon.

In the UK these wrenches are often described by their size, i.e. 18" wrenches are known as "18s", or by the general name of "Stillies/ stills".

Piping and Instrumentation Diagram

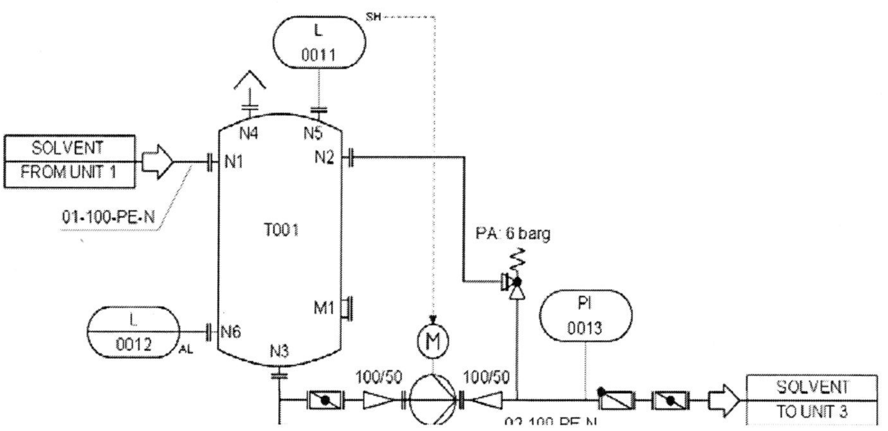

Figure: Piping and instrumentation diagram of pump with storage tank. Symbols according to EN ISO 10628 and EN 62424.

A piping and instrumentation diagram/drawing (P&ID) is a diagram in the process industry which shows the piping of the process flow together with the installed equipment and instrumentation.

Contents and Function

A piping and instrumentation diagram/drawing (P&ID) is defined by the Institute of Instrumentation and Control as follows:

1. A diagram which shows the interconnection of process equipment and the instrumentation used to control the process. In the process industry, a standard set of symbols is used to prepare drawings of processes. The instrument symbols used in these drawings are generally based on International Society of Automation (ISA) Standard S5.
2. The primary schematic drawing used for laying out a process control installation.

P&IDs play a significant role in the maintenance and modification of the process that it describes. It is critical to demonstrate the physical sequence of equipment and systems, as well as how these systems connect. During the design stage, the diagram also provides the basis for the development of system control schemes, allowing for further safety and operational investigations, such as the hazard and operability study (HAZOP).

For processing facilities, it is a pictorial representation of:

- Key piping and instrument details
- Control and shutdown schemes
- Safety and regulatory requirements
- Basic start up and operational information

List of P&ID Items

- Instrumentation and designations
- Mechanical equipment with names and numbers
- All valves and their identifications
- Process piping, sizes and identification
- Miscellanea-vents, drains, special fittings, sampling lines, reducers, increasers and swagers
- Permanent start-up and flush lines
- Flow directions
- Interconnections references
- Control inputs and outputs, interlocks
- Interfaces for class changes
- Computer control system input
- Identification of components and subsystems delivered by others

Identification and Reference Designation

The P&ID is used for the identification of measurements within the process. Identification letters for measurements are based on Standard S5. 1 and ISO 14617-6:

First-Letter	*Measurement*
A	Analysis
B	Burner, Combustion
C	User's Choice (usually Conductivity)
D	User's Choice (usually Density)
E	Voltage
F	Flow
G	User's Choice
H	Hand
I	Current
J	Power
K	Time, Time Schedule
L	Level
M	User's Choice
N	User's Choice (usually Torque)
O	User's Choice
P	Pressure
Q	Quantity
R	Radiation
S	Speed, Frequency
T	Temperature
U	User's Choice (usually Alarm Output)
V	Vibration, Mechanical Analysis
W	Weight, Force
X	User's Choice (usually on-off valve as XV)
Y	Event, State, Presence
Z	Position, Dimension

For reference designation of any equipment in industrial systems the standard IEC 61346 (*Industrial systems, installations and equipment and industrial products — Structuring principles and reference designations*) can be applied. For the function *Measurement* the reference designator B is used, followed by the above listed letter for the measured variable.

For reference designation of any equipment in a power station the KKS Power Plant Classification System can be applied.

Repiping

Repiping means replacing the pipes in a building, oil or gas well, or centrifuge.

Reynolds Number

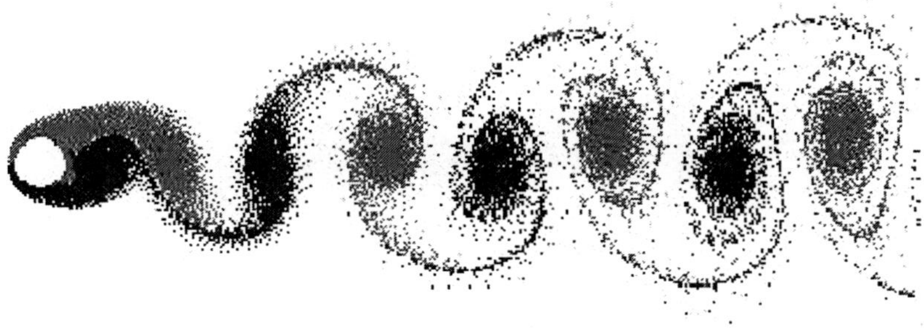

Figure: A vortex street around a cylinder. This occurs around cylinders, for any fluid, cylinder size and fluid speed, provided that there is a Reynolds number of between ~49 and 10000000.

In fluid mechanics, the Reynolds number (Re) is a dimensionless number that gives a measure of the ratio of inertial forces to viscous forces and consequently quantifies the relative importance of these two types of forces for given flow conditions.

The concept was introduced by George Gabriel Stokes in 1851, but the Reynolds number is named after Osborne Reynolds (1842–1912), who popularised its use in 1883. Reynolds numbers frequently arise when performing dimensional analysis of fluid dynamics problems, and as such can be used to determine dynamic similitude between different experimental cases.

They are also used to characterise different flow regimes, such as laminar or turbulent flow: laminar flow occurs at low Reynolds numbers, where viscous forces are dominant, and is characterised by smooth, constant fluid motion; turbulent flow occurs at high Reynolds numbers and is dominated by inertial forces, which tend to produce chaotic eddies, vortices and other flow instabilities.

Definition

Reynolds number can be defined for a number of different situations where a fluid is in relative motion to a surface. These definitions generally include the fluid properties of density and viscosity, plus a velocity and a characteristic length or characteristic dimension. This dimension is a matter of convention – for example a radius or diametre are equally valid for spheres or circles, but one is chosen by convention. For aircraft or ships, the length or width can be used. For flow in a pipe or a sphere moving in a fluid the internal diametre is generally used today. Other shapes such as rectangular pipes or non-spherical objects have an equivalent diametre defined. For fluids of variable density such as compressible gases or fluids of variable viscosity such non-Newtonian fluids, special rules apply. The velocity may also be a matter of convention in some circumstances, notably stirred vessels.

$$\mathrm{Re} = \frac{\rho \mathrm{v} L}{\mu} = \frac{\mathrm{v} L}{\nu}$$

where:

- vis the mean velocity of the object relative to the fluid (SI units: m/s)
- *L*is a characteristic linear dimension, (travelled length of the fluid; hydraulic diametre when dealing with river systems) (m)
- μis the dynamic viscosity of the fluid (Pa ·s or N ·s/m² or kg/(m ·s))
- *V*is the kinematic viscosity ($\mathbf{v} = \mu / \rho$) (m²/s)
- ρis the density of the fluid (kg/m³)

Note that multiplying the Reynolds number by $\frac{vL}{vL}$ yields $\frac{\rho v^2 L^2}{\mu v L}$, which is the ratio of the inertial forces to the viscous forces. It could also be considered the ratio of the total momentum transfer to the molecular momentum transfer.

Flow in Pipe

For flow in a pipe or tube, the Reynolds number is generally defined as:

$$\mathrm{Re} = \frac{\rho \mathrm{v} D_H}{\mu} = \frac{\mathrm{v} D_H}{\nu} = \frac{\mathrm{Q} D_H}{\nu A}$$

where:

- D_H is the hydraulic diametre of the pipe; its characteristic travelled length, *L*, (m).
- *Q*is the volumetric flow rate (m³/s).
- *A*is the pipe *cross-sectional* area (m²).

- υ is the mean velocity of the object relative to the fluid (SI units: m/s).
- μ is the dynamic viscosity of the fluid (Pa. s or N·s/m^2 or kg/(m·s)).
- υ is the kinematic viscosity ($\nu = \mu / \rho$) (m^2/s).
- ρ is the density of the fluid (kg/m^3).

For shapes such as squares, rectangular or annular ducts where the height and width are comparable, the characteristic dimension for internal flow situations is taken to be the hydraulic diametre, D_H, defined as:

$$D_H = \frac{4A}{P}.$$

where A is the cross-sectional area and P is the wetted perimetre. The wetted perimetre for a channel is the total perimetre of all channel walls that are in contact with the flow. This means the length of the water exposed to air is *not* included in the wetted perimetre.

For a circular pipe, the hydraulic diametre is exactly equal to the inside pipe diametre, as can be shown mathematically.

For an annular duct, such as the outer channel in a tube-in-tube heat exchanger, the hydraulic diametre can be shown algebraically to reduce to

$$D_{H,\text{annulus}} = D_o - D_i$$

where

D_o is the inside diametre of the outside pipe, and

D_i is the outside diametre of the inside pipe.

For calculations involving flow in non-circular ducts, the hydraulic diametre can be substituted for the diametre of a circular duct, with reasonable accuracy.

Flow in a Wide Duct

For a fluid moving between two plane parallel surfaces—where the width is much greater than the space between the plates—then the characteristic dimension is twice the distance between the plates.

Flow in an Open Channel

For flow of liquid with a free surface, the hydraulic radius must be determined. This is the cross-sectional area of the channel divided by the wetted perimetre. For a semi-circular channel, it is half the radius. For a rectangular channel, the hydraulic radius is the cross-sectional area divided by the wetted perimetre. Some texts then use a characteristic dimension that is 4 times the hydraulic radius, chosen because it gives the same value of Re for the onset of turbulence as in pipe flow, while others use the hydraulic radius as the characteristic length-scale with consequently different values of Re for transition and turbulent flow.

Object in a Fluid

The Reynolds number for an object in a fluid, called the particle Reynolds number and often denoted Re_p, is important when considering the nature of flow around that grain, whether or not vortex shedding will occur, and its fall velocity.

Sphere in a Fluid

For a sphere in a fluid, the characteristic length-scale is the diametre of the sphere and the characteristic velocity is that of the sphere relative to the fluid some distance away from the sphere, such that the motion of the sphere does not disturb that reference parcel of fluid. The density and viscosity are those belonging to the fluid. Note that purely laminar flow only exists up to Re = 0.1 under this definition. Under the condition of low Re, the relationship between force and speed of motion is given by Stokes' law.

Oblong Object in a Fluid

The equation for an oblong object is identical to that of a sphere, with the object being approximated as an ellipsoid and the axis of length being chosen as the characteristic length scale. Such considerations are important in natural streams, for example, where there are few perfectly spherical grains. For grains in which measurement of each axis is impractical, sieve diametres are used instead as the characteristic particle length-scale. Both approximations alter the values of the critical Reynolds number.

Fall Velocity

The particle Reynolds number is important in determining the fall velocity of a particle. When the particle Reynolds number indicates laminar flow, Stokes' law can be used to calculate its fall velocity. When the particle Reynolds number indicates turbulent flow, a turbulent drag law must be constructed to model the appropriate settling velocity.

Packed Bed

For fluid flow through a bed of approximately spherical particles of diametre D in contact, if the void age is ε and the superficial velocity is V, the Reynolds number can be defined as:

$$\mathrm{Re} = \frac{\rho V D}{\mu(1-\epsilon)}.$$

Laminar conditions apply up to Re = 10, fully turbulent from 2000.

Stirred Vessel

In a cylindrical vessel stirred by a central rotating paddle, turbine or propeller, the characteristic dimension is the diametre of the agitator D. *The* velocity is ND where N is the rotational speed. Then the Reynolds number is:

$$\mathrm{Re} = \frac{\rho N D^2}{\mu}.$$

The system is fully turbulent for values of Re above 10 000.

Transition and Turbulent Flow

In boundary layer flow over a flat plate, experiments confirm that, after a certain length of flow, a laminar boundary layer will become unstable and become turbulent. This instability occurs across different scales and with different fluids, usually when $\mathrm{Re}_x \approx 5\times10^5$, where x is the distance from the leading edge of the flat plate, and the flow velocity is the freestream velocity of the fluid outside the boundary layer. For flow in a pipe of diametre D, experimental observations show that for "fully developed" flow, laminar flow occurs when $\mathrm{Re}_D < 2300$ and turbulent flow occurs when $\mathrm{Re}_D > 4000$. In the interval between 2300 and 4000, laminar and turbulent flows are possible and are called "transition" flows, depending on other factors, such as pipe roughness and flow uniformity. This result is generalised to non-circular channels using the hydraulic diametre, allowing a transition Reynolds number to be calculated for other shapes of channel. These transition Reynolds numbers are also called critical Reynolds numbers, and were studied by Osborne Reynolds around 1895.

Pipe Friction

Pressure drops seen for fully developed flow of fluids through pipes can be predicted using the Moody diagram which plots the

Darcy–Weisbach friction factor *f* against Reynolds number and relative roughness ϵ / D. The diagram clearly shows the laminar, transition, and turbulent flow regimes as Reynolds number increases. The nature of pipe flow is strongly dependent on whether the flow is laminar or turbulent.

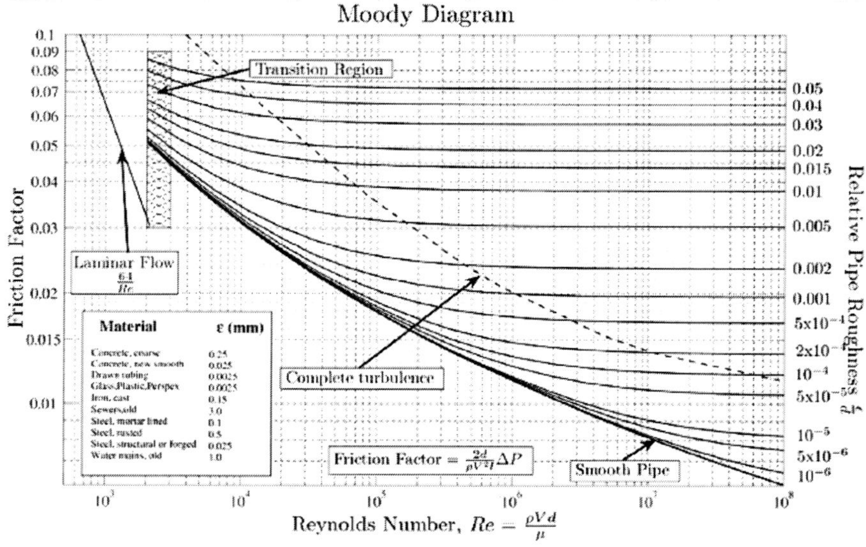

Figure: The Moody diagram, which describes the Darcy–Weisbach friction factor *f* as a function of the Reynolds number and relative pipe roughness.

Similarity of Flows

In order for two flows to be similar they must have the same geometry, and have equal Reynolds numbers and Euler numbers. When comparing fluid beahviour at corresponding points in a model and a full-scale flow, the following holds:

$$\mathrm{Re}_m = \mathrm{Re}$$

$$\mathrm{Eu}_m = \mathrm{Eu} \qquad \text{i.e.} \quad \frac{p_m}{\rho_m {v_m}^2} = \frac{p}{\rho v^2},$$

quantities marked with ‘m’ concern the flow around the model and the others the actual flow. This allows engineers to perform experiments with reduced models in water channels or wind tunnels, and correlate the data to the actual flows, saving on costs during experimentation and on lab time.

Note that true dynamic similitude may require matching other dimensionless numbers as well, such as the Mach number used in compressible flows, or the Froude number that governs open-channel flows. Some flows involve more dimensionless parameters than can be practically satisfied with the available apparatus and fluids, so one is forced to decide which parameters are most important. For experimental flow modelling to be useful, it requires a fair amount of experience and judgement of the engineer.

Smallest Scales of Turbulent Motion

In a turbulent flow, there is a range of scales of the time-varying fluid motion. The size of the largest scales of fluid motion (sometimes called eddies) are set by the overall geometry of the flow. For instance, in an industrial smoke stack, the largest scales of fluid motion are as big as the diametre of the stack itself. The size of the smallest scales is set by the Reynolds number. As the Reynolds number increases, smaller and smaller scales of the flow are visible. In a smoke stack, the smoke may appear to have many very small velocity perturbations or eddies, in addition to large bulky eddies. In this sense, the Reynolds number is an indicator of the range of scales in the flow. The higher the Reynolds number, the greater the range of scales. The largest eddies will always be the same size; the smallest eddies are determined by the Reynolds number.

What is the explanation for this phenomenon? A large Reynolds number indicates that viscous forces are not important at large scales of the flow. With a strong predominance of inertial forces over viscous forces, the largest scales of fluid motion are undamped—there is not enough viscosity to dissipate their motions. The kinetic energy must "cascade" from these large scales to progressively smaller scales until a level is reached for which the scale is small enough for viscosity to become important (that is, viscous forces become of the order of inertial ones). It is at these small scales where the dissipation of energy by viscous action finally takes place. The Reynolds number indicates at what scale this viscous dissipation occurs. Therefore, since the largest eddies are dictated by the flow geometry and the smallest scales are dictated by the viscosity, the Reynolds number can be understood as the ratio of the largest scales of the turbulent motion to the smallest scales.

In Physiology

Poiseuille's law on blood circulation in the body is dependent on laminar flow. In turbulent flow the flow rate is proportional to the square root of the pressure gradient, as opposed to its direct proportionality to pressure gradient in laminar flow.

Rapid changes in vessel diametre may lead to turbulent flow, for instance when a narrower vessel widens to a larger one. Furthermore, a bulge of atheroma may be the cause of turbulent flow, where audible turbulence may be detected with a stethoscope.

In Viscous Fluids

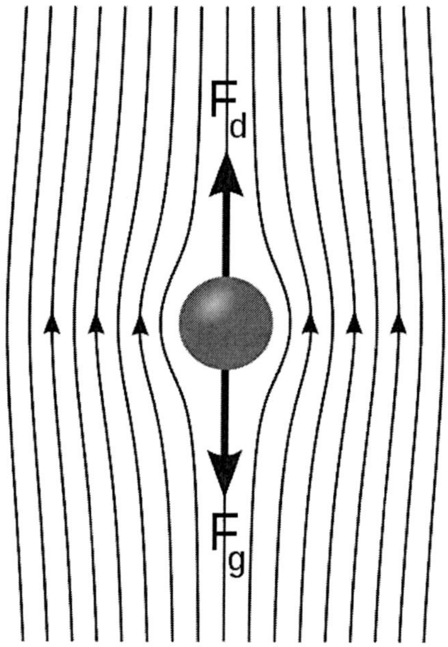

Figure: Creeping flow past a sphere: streamlines, drag force F_d and force by gravity F_g.

Where the viscosity is naturally high, such as polymer solutions and polymer melts, flow is normally laminar. The Reynolds number is very small and Stokes' Law can be used to measure the viscosity of the fluid. Spheres are allowed to fall through the fluid and they reach the terminal velocity quickly, from which the viscosity can be determined.

The laminar flow of polymer solutions is exploited by animals such as fish and dolphins, who exude viscous solutions from their skin to aid flow over their bodies while swimming. It has been used in yacht racing by owners who want to gain a speed advantage by pumping a polymer solution such as low molecular weight polyoxyethylene in water, over the wetted surface of the hull.

It is, however, a problem for mixing of polymers, because turbulence is needed to distribute fine filler (for example) through the material. Inventions such as the "cavity transfer mixer" have been developed to produce multiple folds into a moving melt so as to improve mixing efficiency. The device can be fitted onto extruders to aid mixing.

Derivation

The Reynolds number can be obtained when one uses the nondimensional form of the incompressible Navier-Stokes equations:

$$\rho\left(\frac{\partial \mathbf{v}}{\partial t}+\mathbf{v}\cdot\nabla\mathbf{v}\right)=-\nabla p+\mu\nabla^2\mathbf{v}+\mathbf{f}$$

Each term in the above equation has the units of a "body force" (force per unit volume) or, equivalently, an acceleration times a density. Each term is thus dependent on the exact measurements of a flow. When one renders the equation nondimensional, that is when we multiply it by a factor with inverse units of the base equation, we obtain a form which does not depend directly on the physical sizes. One possible way to obtain a nondimensional equation is to multiply the whole equation by the following factor:

$$\frac{D}{\rho V^2}$$

where:

- V is the mean velocity,v or v, relative to the fluid (m/s).
- D is the characteristic length, L, (m).
- ρ is the fluid density (kg/m^3)

If we now set:

$$\mathbf{v}'=\frac{\mathbf{v}}{V},\ p'=p\frac{1}{\rho V^2},\ \mathbf{f}'=\mathbf{f}\frac{D}{\rho V^2},\ \frac{\partial}{\partial t'}=\frac{D}{V}\frac{\partial}{\partial t},\ \nabla'=D\nabla$$

we can rewrite the Navier-Stokes equation without dimensions:

$$\frac{\partial \mathbf{v}'}{\partial t'}+\mathbf{v}'\cdot\nabla'\mathbf{v}'=-\nabla' p'+\frac{\mu}{\rho DV}\nabla'^2\mathbf{v}'+\mathbf{f}'$$

where the term : $\frac{\mu}{\rho DV}=\frac{1}{\mathrm{Re}}$

Finally, dropping the primes for ease of reading:

$$\frac{\partial \mathbf{v}}{\partial t}+\mathbf{v}\cdot\nabla\mathbf{v}=-\nabla p+\frac{1}{\mathrm{Re}}\nabla^2\mathbf{v}+\mathbf{f}$$

This is why mathematically all flows with the same Reynolds number are comparable. Notice also, in the above equation, as : the viscous terms vanish. Thus, high Reynolds number flows are approximately inviscid in the free-stream.

Riser Clamp

A Riser clamp is a device used by mechanical building trades to support vertical runs of piping at each floor level. The devices are placed around the pipe and integral fasteners are then tightened to clamp them onto the pipe.

The friction between the pipe and riser transfers the weight of the pipe through the riser to the building structure. Risers are generally located at floor penetrations, particularly for continuous floor slabs such as concrete.

They may also be located at some other interval as dictated by local building codes or at intermediate intervals to support plumbing which has been altered or repaired. Heavier piping types such as cast iron require more frequent support. Ordinarily, riser clamps are made of carbon steel and individually sized to fit certain pipe sizes. The term "riser" is typically used to refer to vertical runs of pipes, electrical conduit and tubes.

Sleeve (Construction)

Unfinished, fire-resistance rated drywall assembly with mechanical pipe and sleeve penetration. The plumber has hung the sleeve, then the drywaller propped the sleeve up with drywall, to be followed by mud and tape. The sleeve unnecessarily acts as a further heatsink and is actually a penetrant on its own, with its own exterior gap, requiring firestopping and exposing the firestop sealant on the unexposed side on both bonding surfaces.

Figure: improper sleeving in a drywall assembly.

In construction, a sleeve is used both by the electrical and mechanical trades to create a penetration.

Materials

Sleeves can be made of:

- sections of steel pipe
- plastic
- sheet metal
- proprietary devices that are listed firestop components.

Requirements

- Sleeves must be sized such as to adequately allow the passage of the intended penetrant(s) plus enough room to permit the practical installation and mounting of the penetrants as well as adequate room for firestops. A general practice is to size the sleeve two NPS (pipe sizes) up from the diametre of the penetrant. For example, a 4" pipe, with 1" of thermal insulation makes a 6" penetrant (1" pipe covering on each side of the pipe), plus two pipe sizes = an 8" sleeve, creating a 1" annulus.
- In case of insulated piping, the size of the insulation must be taken into account for the intended firestop certification listing.

Hazards

- Metallic sleeves are heatsinks in the firestop that follows the mounting of the penetrants. Maximum and minimum tolerances for wall thicknesses must be taken into account prior to casting. Heatsinks can affect T-ratings. Organic sealants used for topcaulking in firestops may let go of the sleeve if it has conducted too much heat through to the unexposed side (as in the case of the fire test article, this picture).
- Plastic sleeves are usually removed after the concrete forms are stripped, as they contribute fuel to an accidental fire.

Soluforce

Soluforce is a brandname of Pipelife for a Reinforced Thermoplastic Pipe (RTP). Soluforce is an RTP, which is a flexible pipe, supplied in long length coils up to 400m length and pressure ratings from 30 to 90Bar. In the last few years, this type of pipe has been acknowledged as the standardised solution for oilfield flowlines and also recently used offshore for water injection risers and oil flowlines.

A very great advantage of this pipe is also its very fast installation time compared to steel pipe as speeds up to 1000m/hr have been reached installing RTP in ground surface. The pipe mainly benefits applications where steel fails due to corrosion and installation time is an issue. Beside the application for oil transportation, this pipe is also applied for gas transport lines.

Production

The Soluforce RTP is a three layer pipe construction:

1. A HDPE liner pipe
2. The reinforcement layer, typically Aramid (Twaron or Kevlar)
3. The HDPE protective outer layer for UV, damage and abrasion protection.

Typical Applications

Soluforce is used for the following applications:

- Oil field flowlines
- Oil field waste water disposal lines
- Offshore water injection risers
- Offshore oil flowlines
- High pressure Water transportation lines
- High pressure gas transport lines
- Relining existing pipes.

Testing and Qualification

Solufoce RTP is tested and acknowledged by the following instances:

- DNV Certification D-2615-Soluforce System 4" and 5" with in-line couplings and end fittings
- ASTM-WK11803
- API-RP 15S (oil field service)
- ISO/TS 18226:2006 (gas service)
- DVGW VP 642 (German gas service)
- NYSEARCH project by the Northeast Gas Association (USA)

Sprinkler Fitting

Sprinkler fitting is an occupation consisting of the installing, testing, inspecting, and certifying of automatic fire suppression systems in all types of structures. Sprinkler systems installed by sprinkler fitters can include the underground supply as well as integrated overhead piping systems and standpipes. The fire suppression piping may contain water, air (in a dry system), antifreeze, gas or chemicals as in a hood system, or a mixture producing fire retardant foam.

Figure: Fire sprinkler control valve assembly.

Sprinkler fitters work with a variety of pipe and tubing materials including several types of plastic, copper, steel, cast iron, and ductile iron.

Sprinkler fitters specialise in piping associated with fire sprinkler systems. The piping within these types of systems are required to be installed and maintained in accordance with strict guidelines in order to maintain compliance with the local building code and the fire code. This type of fire protection is considered a part of active fire protection rather than passive fire protection.

Local and National Standards

In the US, fire protection systems must adhere to the standards set forth in the installation standards of NFPA 13, (NFPA) 13D,

(NFPA) 13R, (NFPA 14) and (NFPA) 25 which are administered, copyrighted, and published by the National Fire Protection Association.

Steel Casing Pipe

Steel Casing Pipe, also known as encasement pipe, is most commonly used in underground construction to protect utility lines of various types from getting damaged. Such damage might occur due to the elements of nature or human activity. Steel casing pipe is used in different types of horizontal underground boring, where the pipe is jacked into an augered hole in segments and then connected together by welding or by threaded and coupled ends, or other proprietary pipe connectors. The steel casing pipe can also be set up and welded into a "ribbon" and then directionally pulled through a previously drilled hole under highways, railroads, lakes and rivers.

Uses

Steel casing pipe protects one or many of various types of utilities such as water mains, gas pipes, electrical power cables, fibre optic cables, etc. The utility lines that are run through the steel casing pipe are most commonly mounted and spaced within the steel casing pipe by using "casing spacers" that are made of various materials, including stainless steel or carbon steel and the more economical plastic versions. The ends of a steel casing pipe "run" are normally sealed with "casing end seals", which can be of the "pull-on" or "wrap-around" rubber varieties. Steel casing pipe is also used in the construction of deep foundations.

Specification

Steel casing pipe generally has no specific specifications, other than the need for the material to be extremely straight and round. In some areas A.S.T.M. specifications may be required by project engineers. The specification most commonly called for is A.S.T.M. 139 Grade B. This specification gives parameters for minimum yield and tensile strength of the steel pipe being used for casing, and tolerances of straightness and concentricity. Steel casing pipe is often specified as ASTM A-252 which is a structural grade material that does not require hydrostatic testing and the inspection requirements are not stringent and it usually costs less than other grades such as A-53, A-139 or API 5L. Used natural gas line pipe is also used as casing on many projects because it is often reclaimed in very good condition and can offer a significant cost savings when compared to new steel pipe.

Chapter 5

Thread Protector

A thread protector is used to protect the threads of a pipe during transportation and storage. Thread protectors are generally manufactured from plastic or steel and can be applied to the pipe manually or automatically (by machine).

Thread protectors are used frequently in the oil and gas industry to protect pipes during transportation to the oil and gas fields. Metal thread protectors can be cleaned and re-used, while plastic thread protectors are often collected and either re-used or recycled.

Threaded Pipe

A threaded pipe plug; note that the threaded portion is tapered and so touches the calipers at only two points.

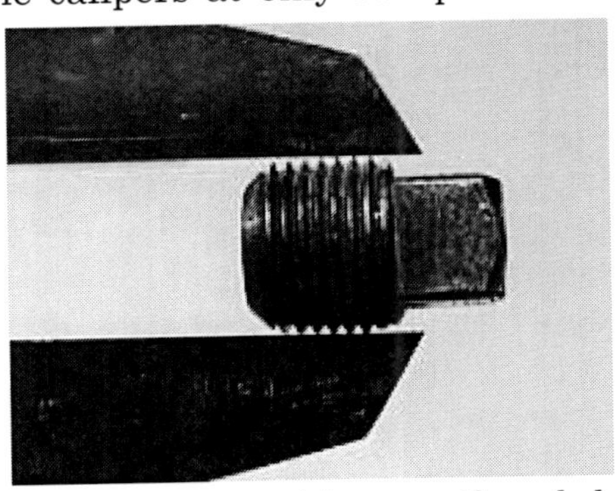

Figure: A threaded pipe is a pipe with screw-threaded ends for assembly.

Tapered Threads

The threaded pipes used in some plumbing installations for the delivery of gases or fluids under pressure have a threaded section that is slightly conical (in contrast to the cylindrical section commonly found on bolts and leadscrews). This is called a "tapered thread". The seal provided by a threaded pipe joint depends upon the labyrinth seal created by the threads; upon a positive seal between the threads created by the deformation of the threads when they are tightened to the proper torque; and sometimes on the presence of a sealing coating, such as thread seal tape ("Teflon tape"), or a liquid or paste pipe sealant such as pipe dope. Tapered thread joints typically do not include a gasket.

Especially precise threads are known as "dry fit" or "dry seal" meaning that no sealant is required for a gas-tight seal. Such threads are needed where the sealant would contaminate or react with the media inside the piping, e.g., oxygen service. Tapered threaded fittings are sometimes used on plastic piping. Due to the wedging effect of the tapered thread, extreme care must be used to avoid over-tightening the joint. The over-stressed female fitting may split days, weeks, or even years after initial installation. Therefore many municipal plumbing codes restrict the use of threaded plastic pipe fittings.

Straight Threads

Pipes may also be threaded with cylindrical threaded sections, in which case the threads do not themselves provide any sealing function other than some labyrinth seal effect, which may not be enough to satisfy either functional or code requirements. With straight pipe threads, the seal may be provided by an O-ring seated between the shoulder of the male pipe section and an interior surface on the female part.

Trace Heating

Electric trace heating, also known as electric heat tracing, heat tape or surface heating, is a system used to maintain or raise the temperature of pipes and vessels. Trace heating takes the form of an electrical heating element run in physical contact along the length of a pipe. The pipe must then be covered with thermal insulation to retain heat losses from the pipe. Heat generated by the element then maintains the temperature of the pipe. Trace heating may be used to protect pipes from freezing, or to maintain process temperatures for piping that must transport substances that solidify at ambient temperatures. Electric trace heating cables are an alternative to steam trace heating where steam is not available or unwanted.

Development

Electric trace heating began in the 1930s but initially no dedicated equipment was available. Mineral insulated cables were run at high current densities to produce heat, and control equipment was adapted from other applications. Mineral-insulated resistance heating cable was introduced in the 1950s, and parallel-type heating cables that could be cut to length in the field became available. Self-limiting thermoplastic cables were marketed in 1971.

Control systems for trace heating systems developed from capillary filled-bulb thermostats and contactors in the 1970s to networked computerised controls in the 1990s, in large systems that require centralised control and monitoring. One paper projected that between 2000 and 2010 trace heating would account for 100 megawatts of connected load, and that trace heating and insulation would account for up to $700 million CDN capital investment in the Alberta oil sands. International standards applied in the design and installation of electric trace heating systems include IEEE standards 515 and 622, British standard BS 6351, and IEC standard 60208.

Uses

The most common pipe trace heating applications include:

- Freeze protection
- Temperature maintenance

Other uses of trace heating cables include

- Ramp and stair snow / ice protection
- Gulley and roof snow / ice protection
- Underfloor heating
- Door / frame interface ice protection
- Window de-misting
- Anti-condensation
- Pond freeze protection
- Soil warming.

Freeze Protection

Every pipe or vessel is subject to heat loss when its temperature is greater than ambient temperature. Thermal insulation reduces the rate of heat loss but does not eliminate it. Trace heating is used to replace the heat that is lost to atmosphere. If the heat replaced matches the heat lost, temperature will be maintained. Normally, a thermostat is used to energise when it measures temperature falling below a set temperature value-usually between 3°C and 5°C and often referred to as the 'setpoint'. The thermostat will de-energise the trace heating when it measures temperature rising past another set temperature value-usually 2°C higher than the setpoint value.

Gutter and Roof De-icing

Placement of heat trace cable on roofs or in gutters to melt ice during winter months. When used in gutters the cable is not meant to keep the gutters free of ice or snow, but only to provide a free path for the melted water to get off the roof and down the downspout or drain piping.

Temperature Maintenance

Hot water service piping can also be traced, so that a circulating system is not needed to provide hot water at outlets. The combination of trace heating and the correct thermal insulation for the operating ambient temperature maintains a thermal balance where the heat output from the trace heating matches the heat loss from the pipe. Self limiting or regulating heating tapes have been developed and are very successful in this application. A similar principle can be applied to process piping carrying fluids which may congeal at low temperatures, for example, tars or molten sulphur. Hit-temperature trace heating elements can prevent blockage of pipes. Industrial applications for trace heating range from chemical industry, oil refineries, nuclear power plants, food factories. For example, wax is a material which starts to solidify below 70 °C which is usually far above the temperature of the surrounding air. Therefore the pipeline must be provided with an external source of heat to prevent the pipe and the material inside it from cooling down. Trace heating can also be done with steam, but this requires a source of steam and may be inconvenient to install and operate. In laboratories, researchers working in the field of materials science use trace heating to heat a sample isotropically. They may use trace heating in conjunction with a variac, so as to control the heat energy delivered. This is an effective means of slowly heating an object to measure thermodynamic properties such as thermal expansion.

Types

Constant Electric Power "Series"

A series heating cable is made of a run of high-resistance wire, insulated and often enclosed in a protective jacket. It is powered at a specific voltage and the resistance of the wire creates heat. The downside of these types of heaters is that if they are crossed over themselves they can overheat and burn out, they are provided in specific lengths and cannot be shortened in the field, also, a break anywhere along the line will result in a failure of the entire cable.

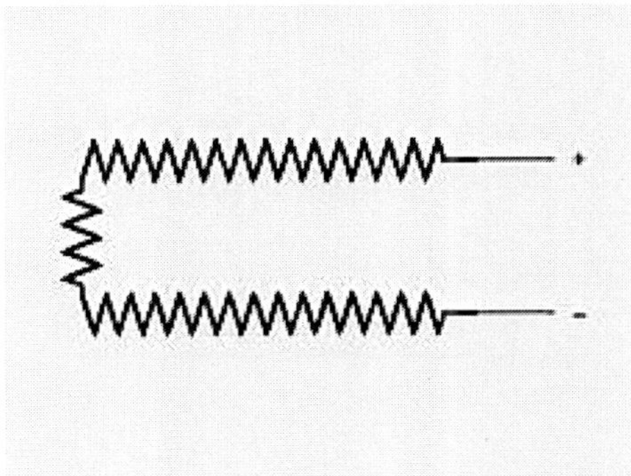

Figure: Series heating Cable

The upside is that they are typically inexpensive (if plastic style heaters) or, as is true with mineral insulated heating cables, they can be exposed to very high temperatures. Mineral insulated heating cables are good for maintaining high temperatures on process lines or maintaining lower temperatures on lines which can get extremely hot such as high temperature steam lines.

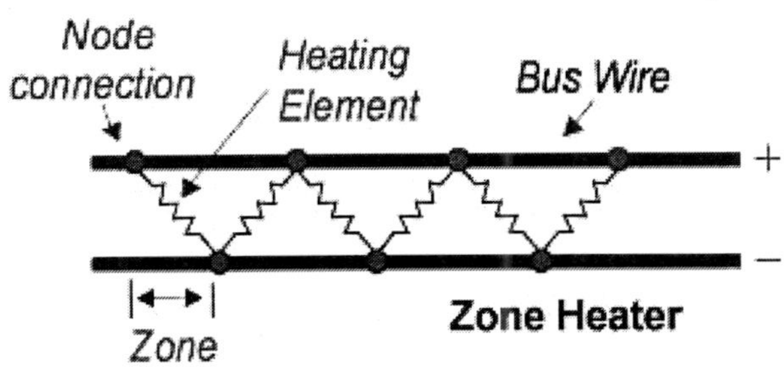

Figure: Typically series elements are used on long pipe line process heating, for example long oil pipe lines and quay side of load pipes on oil refineries.

Constant Electric Power "Zone"

A constant electric power zone cable is made by wrapping a fine heating element around two insulated parallel bus wires, then on alternating sides of the conductors a notch is made in the insulation. The heating element is then normally soldered to the exposed conductor wire which creates a small heating circuit; this is then repeated along the length of the cable. The benefits of this system over series elements is that should one small element fail then the rest of the system will continue to operate, a draw back of this system is that its length is limited to multiples of the notch distance, so when installing on site you normally have to install slightly beyond the end of the pipe work. It is still subject to overheating and burnout if overlapped, but this is generally bad practice to overlap when installing.

Self Regulating

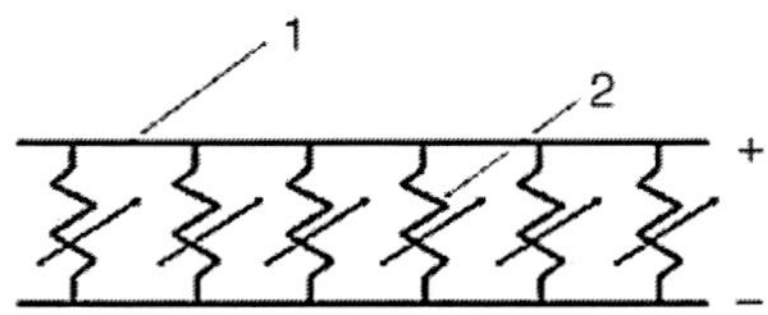

Figure: SR Cable. Bus wires (1), SR polymer (2)

Self-regulating cable uses two parallel bus wires which carry electricity but do not create heat. They are encased in a semi-conductive polymer. This polymer is loaded with carbon; as the polymer element heats, it allows less current to flow.

The cables are manufactured and then irradiated and by varying both the carbon content and the dosage then different tape with different output characteristics can be produced. There is then an inner jacket which separates the bus wires from the grounding braid.

In commercial and industrial cables, an additional outer jacket of rubber or Teflon is applied. The benefits of this cable are the ability to cut to length in the field, It is more rugged but not necessarily more reliable than series or zone heaters, it cannot over-heat itself so in theory it can be crossed, but it is bad practice to install tape in this way. Self regulating heating cables have a specific maximum exposure temperature based on the type of polymer which is used to make the heating core which means that if they are subject to high temperatures then the tape can be damaged beyond repair. Also self limiting tapes are subject to high inrush currents on starting up similar to 'induction' motor so a higher rated contactor is required.

Power Supply and Control

Trace heat cables may be connected to single-phase or (in groups) to three-phase power supplies. Power is controlled either by a contactor or a solid-state controller. For self-regulating cable, the supply must furnish a large warm-up current if the system is switched on from a cold starting condition. The contactor or controller may include a thermostat if accurate temperature maintenance is required, or may just shut off a freeze-protection system in mild weather.

Electrical heat tracing systems may be required to have earth leakage (ground fault or RCD) devices for personnel and equipment protection. The system design must minimise leakage current to prevent nuisance tripping; this may limit the length of any individual heating circuit.

Control System

The three phase systems are fed via contactors similar to a three phase motor 'direct on line' starter which is controlled by a thermostat somewhere in the line. This ensures that the temperature is kept constant and the line does not overheat or underheat.

If a line becomes frozen because the heating was switched off then this may take some time to thaw out using trace heating. This thawing out is done on the three phase systems by using an 'auto transformer' to give a higher voltage, and consequently higher current, and make the trace heating elements a bit hotter. The boost system is usually on a timer and switches back to 'normal' after a period of time.

Tube (Fluid Conveyance)

A tube, or tubing, is a long hollow cylinder used to convey fluids (liquids or gases). The terms "pipe" and "tube" are almost interchangeable, although minor distinctions exist — generally, a tube has tighter engineering requirements than a pipe. Both pipe and tube imply a level of rigidity and permanence, whereas a hose is usually portable and flexible. A tube and pipe may be specified by standard pipe size designations, *e.g.*, nominal pipe size, or by nominal outside or inside diametre and/or wall thickness. The actual dimensions of pipe are usually not the nominal dimensions: A 1-inch pipe will not actually measure 1 inch in either outside or inside diametre, whereas many types of tubing are specified by actual inside diametre, outside diametre, or wall thickness.

Manufacture

There are three classes of manufactured tubing: seamless, as-welded or electric resistant welded (ERW), and drawn-over-mandrel (DOM).

- Seamless tubing is produced via extrusion or rotary piercing.
- Drawn-over-mandrel tubing is made from cold-drawn electrical-resistance-welded tube that is drawn through a die and over a mandrel to create such characteristics as dependable weld integrity, dimensional accuracy, and an excellent surface finish.

Standards

There are many industry and government standards for pipe and tubing. Many standards exist for tube manufacture; some of the most common are as follows:

- ASTM A213 Standard Specification for Seamless Ferritic and Austenitic Alloy-Steel Boiler, Superheater, Heat-Exchanger Tubes.
- ASTM A269 Standard Specification for Seamless and Welded Austenitic Stainless Steel Tubing for General Service
- ASTM A270 Standard Specification for Seamless and Welded Austenitic Stainless Steel Sanitary Tubing
- ASTM A511 Standard Specification for Seamless Stainless Steel Mechanical Tubing
- ASTM A513 Standard Specification for Electric-Resistance-Welded Carbon and Alloy Steel Mechanical Tubing
- ASTM A554 Standard Specification for Welded Stainless Steel Mechanical Tubing
- British Standard 1387:1985 Specification for screwed and socketed steel tubes and tubulars and for plain end steel tubes suitable for welding or for screwing to BS 21 pipe threads

ASTM material specifications generally cover a variety of grades or types that indicate a specific material composition. Some of the most commonly used are:

- TP 304
- TP 316
- MT 304
- MT 403
- MT 506

In installations using hydrogen, copper and stainless steel tubing must be factory pre-cleaned (ASTM B 280) and/or certified as instrument grade. This is due to hydrogen's particular propensities: to explode in the presence of oxygen, oxygenation sources, or contaminants; to leak due to its atomic size; and to cause embrittlement of metals, particularly under pressure.

Calculation of Strength

For a tube of silicone rubber with a tensile strength of 10 MPa and a 8 mm outer diametre and 2 mm thick walls. The maximum pressure may be calculated as follows:

Outer diametre = 0.008 [metre]

Wall thickness = 0.002 [metre]

Tensile strength = 10 * 1000000 [Pa]

Pressure burst = (Tensile strength * Wall thickness * 2 / (10 * Outer diametre)) * 10 [Pa]

Gives burst pressure of 5 MPa.

Using a safety factor:

Pressure max = (Tensile strength * Wall thickness * 2 / (10 * Outer diametre)) * 10 / Safety_factor [Pa]

Tube Bending

Figure: Bent tubing

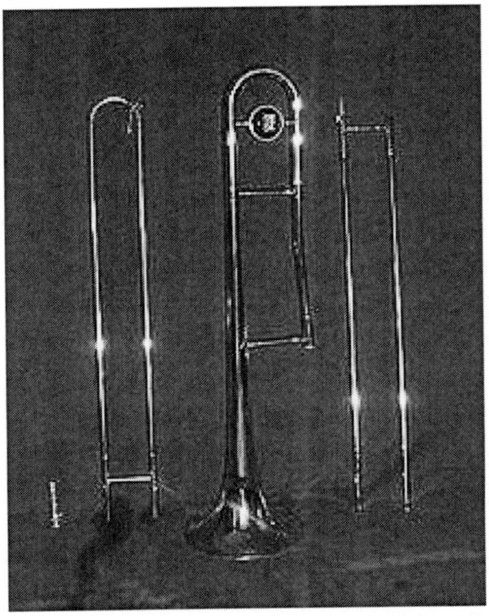

Figure: Shown is a trombone, it displays some U-bends

Tube bending is the umbrella term for metal forming processes used to permanently form pipes or tubing. One has to differentiate between form-bound and freeform-bending procedures, as well as between heat supported and cold forming procedures.

Form bound bending procedures like "press bending" or "rotary draw bending" are used to form the work piece into the shape of a die. Straight tube stock can be formed using a bending machine to create a variety of single or multiple bends and to shape the piece into the desired form. This processes can be used to form complex shapes out of different types of ductile metal tubing. Freeform-bending processes, like *three-roll-pushbending*, shape the workpiece kinematically, thus the bending contour is not dependent on the tool geometry.

Generally, round stock is what is used in tube bending. However, square and rectangular tubes and pipes may also be bent to meet job specifications. Other factors involved in the tube bending process is the wall thickness, tooling and lubricants needed by the pipe and tube bender to best shape the material.

Geometry

A tube can be bent in multiple directions and angles. Common simple bends consist of forming elbows, which are bends that range from 2 to 90°, and U-bends, which are 180° bends. More complex geometries include multiple two-dimensional (2D) bends and three-dimensional (3D) bends. A 2D tube has the openings on the same plane; a 3D has openings on different planes.

A two plane bend or compound bend is defined as a compound bend that has a bend in the plan view and a bend in the elevation. When calculating a 2 plane bend you must know the bend angle and rotation (dihedral angle). One side effect of bending the workpiece is the wall thickness changes; the wall along the inner radius of the tube becomes thicker and the outer wall becomes thinner. To overcome this the tube may be supported internally or externally to preserve the cross section. Depending on the bend angle, wall thickness, and bending process the inside of the wall may wrinkle.

Processes

Tube bending as a process starts with loading a tube into a pipe bender and clamping it into place between two dies, the clamping block and the forming die. The tube is also loosely held by two other dies, the wiper die and the pressure die.

The process of tube bending involves using mechanical force to push stock material pipe or tubing against a die, forcing the pipe or tube to conform to the shape of the die. Often, stock tubing is held firmly in place while the end is rotated and rolled around the die. Other forms of processing including pushing stock through rollers that bend it into a simple curve. For some tube bending processing, a mandrel is placed inside the tube to prevent collapsing. The tube is also held in tension by a wiper die to prevent any creasing during stress. A wiper die is usually made of a softer alloy i.e. aluminium, brass to avoid scratching or damaging the material being bent.

Much of the tooling is made of hardened steel or tooled steel to maintain and prolong the tools life. However wherever there is a concern of scratching or gouging the work piece, a softer material such as aluminium or bronze is utilised. For example, the clamping block, rotating form block and pressure die are often formed from the hardened steel because the tubing is not moving past these parts of the machine. On the other hand, the pressure die and the wiping die are formed from aluminium or bronze to maintain the shape and surface of the workpiece as it slides by.

Pipe bending machines are typically human powered, pneumatic powered, hydraulic assisted, hydraulic driven, or electric servomotor.

Press Bending

Probably will be the first bending process used on cold pipes and tubing. In this process a die in the shape of the bend is pressed against the pipe forcing the pipe to fit the shape of the bend. Because the pipe is not supported internally there is some deformation of the shape of the pipe giving an ovular cross section. This process is used where a consistent cross section of the pipe is not required. Although a single die can produce various shapes, it only works for one size tube and radius.

Rotary Draw Bending

Rotary draw bending (RDB) are precise in that they bend using tooling or "die sets" which have a constant centre line radius (CLR). The die set consists of two parts: The *former die* creates the shape to which the material will be bent. The *counter die* does the work of pushing the material into the former die while travelling the length of the bend. Rotary draw benders can be programmable to store multiple bend jobs with varying degrees of bending. Often a positioning index table (IDX) is attached to the bender allowing the operator to reproduce complex bends which can have multiple bends and differing planes. Rotary draw benders are the most popular machines for use in bending tube, pipe and solids for applications like: handrails, frames, motor vehicle roll cages, handles, lines and much more. Rotary draw benders create aesthetically pleasing bends when the right tooling is matched to the application.

Three-Roll Pushbending

The *Three-Roll Pushbending* (TRPB) is the most commonly used freeform-bending process to manufacture bending geometries consisting of several plain bending curves. Nevertheless, a 3D-shaping is possible. The profile is guided between bending-roll and supporting-roll(s), while being pushed through the tools. The position of the forming-roll defines the bending radius. The bending point is the tangent-point between tube and bending-roll. To change the bending plane, the pusher rotates the tube around its longitudinal axis. Generally, a TRPB tool kit can be applied on a conventional rotary draw bending machine. Bending contours defined as spline- or polynomial-functions can be manufactured.

Heat-induction

An induction coil is placed around a small section of the pipe at the bend point. It is then heated to between 800 and 2,200 degrees Fahrenheit (430 and 1,200 °C). While the pipe is hot, pressure is placed on the pipe to bend it. The pipe is then quenched with either air or water spray. Heat-Induction bending is used on large pipes such as freeway signs, power plants, and petroleum pipe lines.

Roll Benders

During the roll bending process the pipe, extrusion, or solid is passed through a series of rollers (typically 3) that apply pressure to the pipe gradually changing the bend radius in the pipe. The pyramid style roll benders have one moving roll, usually the top roll. Double pinch type roll benders have two adjustable rolls, usually the bottom rolls, and a fixed top roll. This method of bending causes very little deformation in the cross section of the pipe. This process is suited to producing coils of pipe as well as long gentle bends like those used in truss systems.

Sand-packing/hot-slab Forming

In the sand packing process the pipe is filled with fine sand and the ends are capped. The pipe is then heated in a furnace to 1,600 °F (870 °C) or higher. The pipe is then placed on a slab with pins set in it. The pipe is then bent around the pins using a winch, crane, or some other mechanical force. The sand in the pipe minimises distortion in the pipe cross section.

Mandrels

A mandrel is a steel rod or linked ball inserted into the tube while it is being bent to give the tube extra support to reduce wrinkling and breaking the tube during this process. The different types of mandrels are as follows.

1. Plug mandrel, a solid rod used on normal bends.
2. Form mandrel, a solid rod with curved end used on bend when more support is need.
3. Ball mandrel without cable, unlinked steel ball bearings inserted into tube, used on critical and precise bends.
4. Ball mandrel with cable, linked ball bearings inserted into tube, used on critical bend and precise bends.
5. Sand, sand packed into tube.

In production of a product where the bend is not critical a plug mandrel can be used. A form type tapers the end of the mandrel to provide more support in the bend of the tube. When precise bending is needed a ball mandrel (or ball mandrel with steel cable) should be used. The conjoined ball-like disks are inserted into the tubing to allow for bending while maintaining the same diametre throughout. Other styles include using sand, cerrobend, or frozen water. These allow for a somewhat constant diametre while providing an inexpensive alternative to the aforementioned styles.

Bending Springs

These are strong but flexible springs inserted into a pipe to support the pipe walls during manual bending. They have diametres only slightly less than the internal diametre of the pipe to be bent. They are only suitable for bending 15-and-22 mm (0.6-and-0.9 in) soft copper pipe (typically used in household plumbing) or PVC pipe.

The spring is pushed into the pipe until its centre is roughly where the bend is to be. A length of flexible wire can be attached to the end of the spring to facilitate its removal. The pipe is generally held against the flexed knee, and the ends of the pipe are pulled up to create the bend. To make it easier to retrieve the spring from the pipe, it is a good idea to bend the pipe slightly more than required, and then slacken it off a little. They are less cumbersome than rotary benders, but are not suitable for bending short lengths of piping when it is difficult to get the required leverage on the pipe ends.

Bending springs for smaller diametre pipes (10mm copper pipe) slide over the pipe instead of inside.

Water Pipe

Figure: A system of copper water tubes used in a radiator heating system.

Water pipes are pipes or tubes, frequently made of polyvinyl chloride (PVC/uPVC), ductile iron, steel, cast iron, polypropylene, polyethylene, or copper, that carry pressurised and treated fresh water to buildings (as part of a municipal water system), as well as inside the building.

History

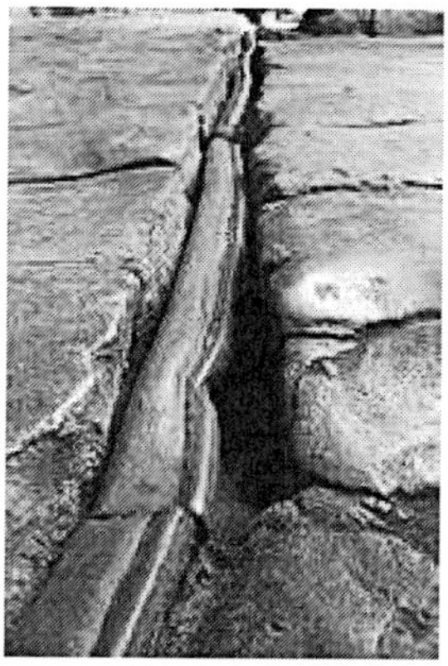

Figure: An original Roman lead pipe with a folded seam, at the Roman Baths in Bath, UK

Figure: Old water pipe, remnant of the Machine de Marly near Versailles, France

For many centuries, lead was the favoured material for water pipes, due to its malleability (this use was so common that the word "plumbing" derives from the Latin word for lead). This was a source of lead-related health problems in the years before the health hazards of ingesting lead were fully understood; among these were stillbirths and high rates of infant mortality. Lead water pipes were still in common use in the early 20th century and remain in many households. Lead-tin alloy solder was commonly used to join copper pipes, but modern practice uses tin-antimony alloy solder to join copper in order to eliminate lead hazards.

Despite the Romans common use of lead pipes, their aqueducts rarely poisoned people. Unlike other parts of the world where lead pipes cause poisoning, the Roman water had so much calcium in it, that a layer of plaque prevented the water contacting the lead itself. What often causes confusion is the large amount of evidence of widespread lead poisoning, particularly amongst those who would have had easy access to piped water. This was an unfortunate result of lead being used in cookware and as an additive to processed food and drink, such as a preservative in wine. Roman lead pipe inscriptions provided information on the owner to prevent water theft.

Wooden pipes were used in London and elsewhere during the 16th and 17th centuries. The pipes were hollowed-out logs, which were tapered at the end with a small hole in which the water would pass through. The multiple pipes were then sealed together with hot animal fat. They were often used in Montreal and Boston in the 1800s, and built-up wooden tubes were widely used in the USA during the 20th century. These pipes, used in place of corrugated iron or reinforced concrete pipes, were made of sections cut from short lengths of wood. Locking of adjacent rings with hardwood dowel pins produced a flexible structure. About 100,000 feet of these wooden pipes were installed during WW2 in drainage culverts, storm sewers and conduits, under highways and at army camps, naval stations, airfields and ordnance plants. Cast iron and ductile iron pipe was long a lower-cost alternative to copper, before the advent of durable plastic materials but special non-conductive fittings must be used where transitions are to be made to other metallic pipes, except for terminal fittings, in order to avoid corrosion owing to electrochemical reactions between dissimilar metals.

Bronze fittings and short pipe segments are commonly used in combination with various materials.

Pipe vs. Tube

Figure: A plastic water pipe being installed. Note that the inner tube is actually transporting the water, while the outer tube only serves as a protective casing

The difference between pipe and tube is simply in the way it is sized. PVC pipe for plumbing applications and galvanised steel pipe for instance, are measured in IPS (iron pipe size). Copper tube, CPVC, PeX and other tubing is measured nominally, which is basically an average diametre.

The thicknesses of the water pipe and tube walls can vary. Pipe wall thickness is denoted by various schedules or for large bore polyethylene pipe in the UK by the Standard Dimension Ratio (SDR) being the ratio of the pipe diametre to its wall thickness. Pipe wall thickness increases with schedule, and is available in schedules 20, 40, 80, and higher in special cases. The schedule is largely determined by the operating pressure of the system, with higher pressures commanding greater thickness. Copper tubing is available in four wall thinkest wang type DWV (thinnest wall; only allowed as drain pipe per UPC), type 'M' (thin; typically only allowed as drain pipe by IPC code), type 'L' (thicker, standard duty for water lines and water service), and type 'K' (thickest, typically used underground between the main and the metre). Because piping and tubing are commodities, having a greater wall thickness implies higher initial cost. Thicker walled pipe generally implies greater durability and higher pressure tolerances.

Wall thickness does not affect pipe or tubing size. 1/2" L copper has the same outer diametre as 1/2" K or M copper. The same applies to pipe schedules. As a result, a slight increase in pressure losses is realised due to a decrease in flowpath as wall thickness is increased. In other words, 1 foot of 1/2" L copper has slightly less volume than 1 foot of 1/2 M copper.

Demand for copper products have fallen due to the dramatic increase in the price of copper, resulting in increased demand for alternative products including PEX and stainless steel.

Weld-On

Weld-On is a division of IPS Corporation, a manufacturer of solvent cements, primers and cleaners for PVC, CPVC, and ABS plastic piping systems. Weld-On products are commonly used for joining plastic pipes and fittings. Weld-On also manufactures specialty products from repair adhesives for leaking pipes, pipe thread sealants / joint compounds, to test plugs for pipeline pressure testing.

Headquartered in California, Weld-On has operations throughout the United States, as well as China, and a worldwide network of sale representatives and distributors.

History

In 1958, the IPS Corporation began endorsing the solvent welding technique, a type of plastic welding, and patented Weld-On solvent cement for use on plastic pipe and fittings. Then in 1978 Weld-On developed colour-match acrylic adhesive for bonding solid surface counter-tops. In 1992 Weld-On formulated and introduced the first low volatile organic compound (VOC) emission solvent cement in response to growing air quality concerns. Five years later Weld-On 724 was introduced as the first high strength solvent cement for chemical resistant CPVC plastic joints in the market and formulated for use in a variety of harsh chemical applications such as hypochlorites, acids and caustics. In 1999, Weld-On began offering environmentally-responsible, all Low VOC solvent cements, primers and cleaners, and began phasing out regular VOC products.

Trap (Plumbing)

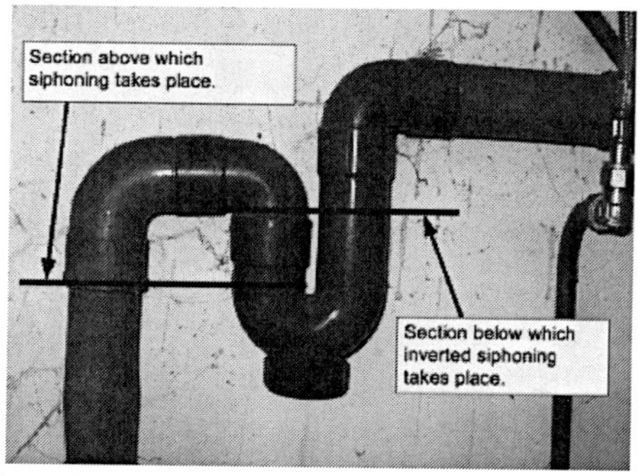

Figure: Water seal under a sink. Siphoning takes place

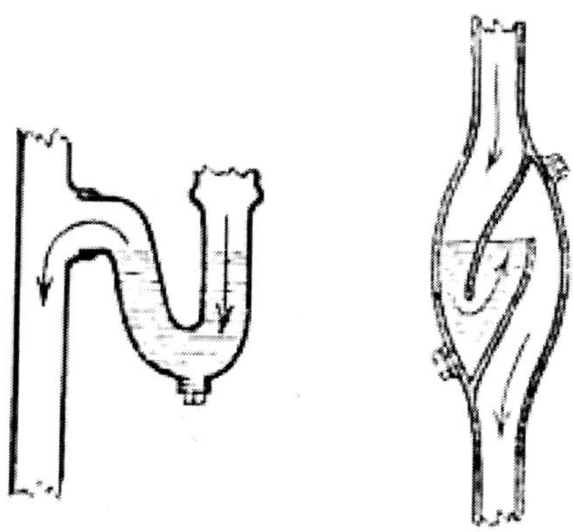

Figure: Examples of traps

In plumbing, a trap is a U-, S-, or J-shaped pipe located below or within a plumbing fixture. An S-shaped trap is also known as the S-bend invented by Alexander Cummings in 1775 but became known as the U-bend following the introduction of the U-shaped trap by Thomas Crapper in 1880. The new U-bend could not jam, so, unlike the S-bend, it did not need an overflow. The bend is used to prevent sewer gases from entering buildings. In refinery applications, it also prevents hydrocarbons and other dangerous gases from escaping outside through drains. The most common of these traps in houses is referred to as a P-trap. It is the addition of a 90 degree fitting on the outlet side of a U-bend, thereby creating a P-like shape. It can also be referred to as a sink trap due to the fact it is installed under most house sinks.

Because of its shape, the trap retains a small amount of water after the fixture's use. This water in the trap creates a seal that prevents sewer gas from passing from the drain pipes back into the occupied space of the building. Essentially all plumbing fixtures including sinks, bathtubs, and toilets must be equipped with either an internal or external trap.

Because it is a localised low-point in the plumbing, sink traps also tend to capture heavy objects (such as jewellery) that are inadvertently dropped into the sink. Traps also tend to collect hair, sand, and other debris and limit the ultimate size of objects that will pass on into the rest of the plumbing, thereby catching over-sized objects. For all of these reasons, most traps can either be disassembled for cleaning or they provide some sort of cleanout feature.

Venting and Auxiliary Devices

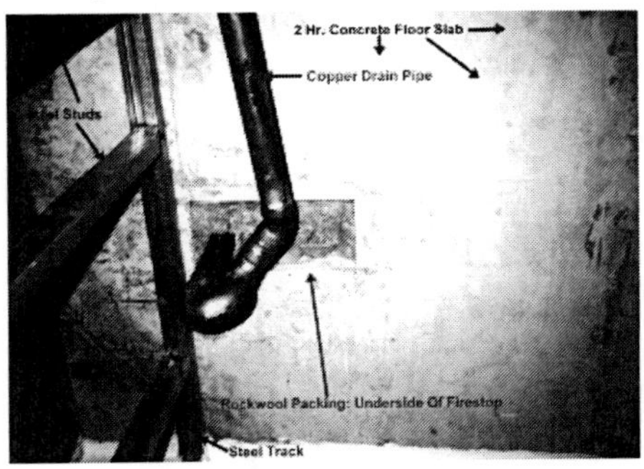

Figure: Trap with copper drain pipe at underside of firestop packing in 2 hour fire-resistance rated concrete floor slab

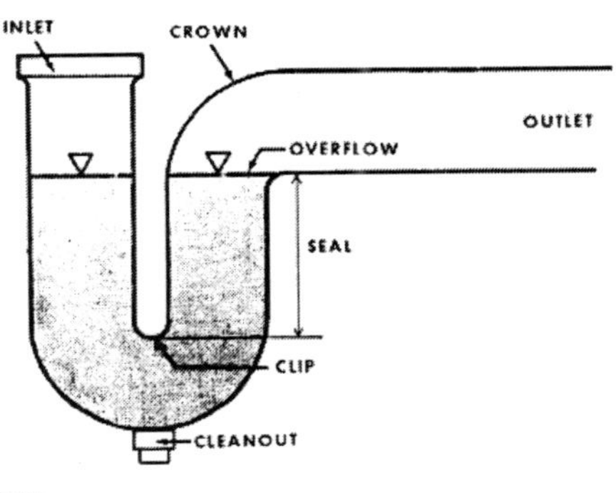

Figure: Typical P-trap

Maintaining the water seal is critical to trap operation; traps can and do dry out, and poor venting can siphon or blow water out of the traps. This is usually avoided by venting the drain pipes downstream of the trap; by being vented to the atmosphere outside the building, the drain lines never operate at a pressure much higher or lower than atmospheric pressure. Plumbing codes usually provide strict limitations on how far a trap may be located from the nearest *vent stack*.

When a vent cannot be provided, codes may allow the use of an air admittance valve instead. These devices avoid negative pressure in the drain pipe by venting room air into the drain pipe (behind the trap). A "Chicago Loop" is another alternative.

When a trap is installed on a fixture that is not routinely used, the eventual evapouration of the water in the trap must be considered. In these cases, a trap primer may be installed; these are devices that automatically recharge traps with water to maintain their water seals.

Accepted Traps

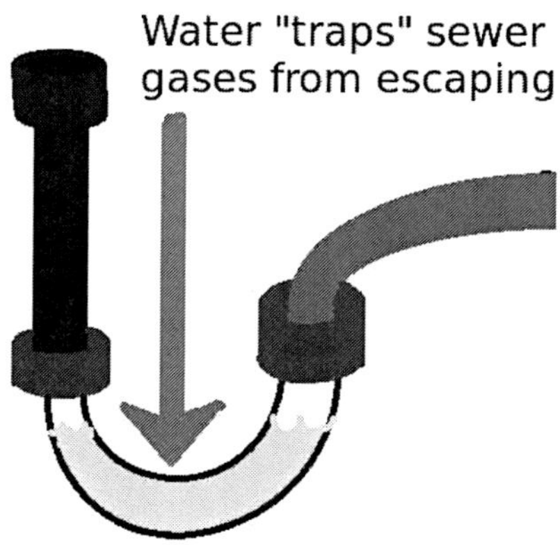

Figure: Gravity holds water to act as a barrier, preventing the escape of sewer gases. If a plumbing fixture is not used over a month or so, the water in the trap will evapourate, allowing noxious odours to escape.

In many locations, "S" traps are no longer accepted by the plumbing codes and are even illegal, as these traps tend to easily siphon dry even when well-vented. It is often possible to tell whether a household uses an S or U-bend by checking for the presence of an overflow pipe outlet. Certain drum-styled traps are also discouraged or banned.

Panzergewinde

The *Stahlpanzerrohrgewinde* standard for screw threads, more often called by the shorter name *Panzergewinde* (pronunciation), is a technical standard created in Germany and subsequently used in various countries, such as the German-speaking zone (Germany, Switzerland, Austria) and neighbouring European countries. The German name "*Stahlpanzerrohrgewinde*" translates to English as 'steel conduit thread'. The thread is used to join pieces of electrical conduit and cable glands. Alternate stylings of the German name are *Stahl-Panzer-Rohr-Gewinde* and an abbreviated form, *StaPa-Rohr-Gewinde.*

The standard, codified by the *Deutsches Institut für Normung* (DIN, German Institute for Standardisation), is DIN 40430. *Panzergewinde* sizes are named with the prefix *PG* plus a nominal number which approximately corresponds to the maximum cable diametre (in millimetres) that can be passed through the conduit.

Because the walls of the conduit are usually relatively thin, the thread depth should not be very large. Thus a thread angle of 80° is used. The *Verband der Elektrotechnik, Elektronik und Informationstechnik* (VDE) (which began as a trade association for standardisation in electrical engineering) originally standardised (and named) the thread for use with conduit and cable glands that were made of steel, although today the thread is used with both steel (typically plated with combinations of nickel, zinc, or tin to resist rusting) and polyvinyl chloride (PVC).

Beginning in 2000, the VDE standard for cable glands (VDE 0619) was formally replaced by EN 50262. After a transitional period of several years during which it could still be used, it was replaced in 2003 by a final metric fine thread with 1.5 mm pitch. Similary, conduit threads were replaced by EN 60423.

Even today, *Panzergewinde* cable glands are still often found on chemical reactors and bioreactors (for example, PG13.5 thread for screwing in sensors) and various other equipment, enclosures, junction boxes, and connectors.

Pronunciation

- *Stahlpanzerrohrgewinde*: SHTAHL-pahntser-rôr-guh-v-n-duh
- *Panzergewinde*: PAHNTSER-guh-v-n-duh
- *PG:* peh-geh

Corrosion in Pipes

Erosion is the gradual destruction of a matter by its environs. Dissimilar metal contact, oxygen, water and debris may cause erosion in plumbing pipes. Left untreated, eroded pipes will cause a range of expensive plumbing difficulties such as bust pipes, leaking slab, and clogged drains. With the yearlong expenditure of erosion estimated at beyond 276 billion dollars-no wonder Americans spend millions of dollars on pipeline corrosion repair in their houses and offices.

By comprehending the crucial facts of home plumbing erosion brought to you by New York Plumber, you may manage or avoid this destructive power to save funds and preserve a healthy environment for your family unit.

Water Pipeline Corrosion: In more than a few circumstances iron, copper and steel pipes and fittings are exposed to erosion destruction. The connection of dissimilar metals such as copper and aluminium or copper and steel causes erosion in pipeline. Metal pipeline erosion is also known as galvanic erosion and is the aftereffect of electrochemical electron swap-over from the differential galvanic properties separating dissimilar metals. Plumbing contractors manage or avoid water pipeline erosion by putting into use anti erosion pipes or at least the matching kinds of metals for plumbing fittings. If erosion protected pipes are not available and plumbing contractors cannot employ matching metals-they will install pipeline erosion protection in the shape of a plastic or fibre connector separating the fittings to avoid dissimilar metals from contacting. The subsequent variables impact erosion in metal pipeline:

- Low pH water values
- Chemical makeup of water
- Water pressure in pipeline
- Quantity of oxygen in water
- Water temperature
- Galvanic erosion

Copper Pipeline Erosion: Pitting erosion causes small leakages in copper water pipes. Copper pipeline erosion causes the formation of small holes on limited patches of the metal shell-causing small leakages that lose beyond seventy water gallons each day. Although copper pipeline erosion develops for unidentified reasons, plenty of plumbing contractors attribute the water chemistry passing through copper pipes to the formation of holes.

Pipeline Erosion Protection: aside from causing more than a few plumbing difficulties, eroded pipes transform water passing through them highly acidic, non-potable, and of a coppery-brown shade. Consequently pipeline erosion repairs have to be carried out once the symptoms for erosion are noticed by you. Tattered or faded patches along pipes or lower than pH seven water values are familiar giveaways for pipeline erosion. Plumbing contractors guide landowners to carry out periodical examinations of exposed copper pipeline for bluish-green spots that in plenty of circumstances indicate small leakages. If noticed, landowners and re-enters have to contact their local plumbing contractors for efficient management. Placing pipe vises on small holes will end leakages and damage control the disaster until certified repair is carried out.

Hydraulic Theory

This section summarises the hydraulic principles which are used in MIDUSS for the analysis and design of pipes. Flow is assumed to be uniform within each reach of pipe, so that the depth and other cross-sectional properties are constant along the length of the pipe. It follows that the bed slope *S0*, the water surface and the slope of the energy line *Sf* are all parallel. The resistance is assumed to be represented by the Manning equation:

$$Q = \frac{M}{n} A R^{2/3} S_o^{\ 1/2}$$

where

Q = normal discharge (c.m/s or c.ft/s)

M =1.0 for metric units

1.49 for imperial or US customary units

n =Manning's roughness coefficient

A =cross-sectional area

R =hydraulic radius = Area/Wetted perimetre

S0 =bed slope (m/m or ft/ft)

No allowance is made for any apparent variation of 'n' with the relative depth of flow in the pipe.

Piping Specs and Safety

The pressure ratings I posted yesterday came from a reference commonly used in the hydraulic industry and were stated for ASTM A53 welded and seamless pipe. Looking in other references and texts today, I find different burst pressures for seamless pipe. Since the majority of the references call for the higher pressures for seamless, I will post those here. Each person must decide what factor of safety they want to use. What I do for myself may not be the same as what you will want to do. There are MANY factors that affect the pressure rating of pipe. I suggest, as I did yesterday, that an informed individual uses at least a 6 to 1 factor of safety on burst pressure for properly installed new pipe. Below are the numbers that have two good references in agreement. These would be 6 to 1 on the stated burst pressure for ASTM A53 grade B SEAMLESS and ASTM A106 grade B which is also seamless. Note that welded pipe will have lower pressure capability.

1/2" ASTM A53 grade B type S and ASTM A106 grade B schedule 40 2600 psi

3/4" ASTM A53 grade B type S and ASTM A106 grade B schedule 40 2150 psi

1/2" ASTM A53 grade B type S and ASTM A106 grade B schedule 80 3500 psi

3/4" ASTM A53 grade B type S and ASTM A106 grade B schedule 80 2933 psi

All good pipe is identified on the outside of the pipe. Usually with a paint or ink marking. ASTM is the most recognised pipe spec. Any pipe not identified on the outside or not readable should NOT be used for hydraulics. There are pipe size tubings out there that are real junk and will not hold pressure. Some are sold in home centres as pipe.

ASTM A53 covers both welded and seamless black and galvanised pipe. ASTM A106 is only seamless and usually for high temp service, but it can be used for hydraulic systems. The grade identifies the minimum strength of the pipe material. You want at least grade B. Grade A is lower strength.

There are also API specs. API 5/L Grade B is very similar to A53 Grade B type S. API 5/LX is stronger and better.

Pipe may be Rated and Marked for more than One Spec.

Again, you must educate yourself. Don't in any way think that you can say "Well I found this on the internet and this must be OK" You can't do that because you have not explained in detail how you are doing other things that have a very big impact on the pressure rating of the pipe. The list is very long.

A safety note on hydraulic presses used to forge. Please remember that a pin hole leak of hydraulic oil if it impacts the hot metal will make a flame-thrower. Please use a fire resistant fluid, such as a water/glycol, or polygycol fluid. These fluids, if maintained at the correct water content are very fire resistant. Please consider the hoses on the machine and replace at any signs of damage. It is a good idea to avoid hoses and instead pipe with a good grade of hydraulic pipe such as seamless schedule 80. I work in the forging industry, and have for 24 years. For 21 years of that I worked for a valve and fitting maker that supplied the heavy hydraulic press trade. I have seen the results of pinhole leaks in oil systems on forging equipment, and it is not pretty. Please note that in my previous post I suggested schedule 80 seamless, and in past posts forged steel fittings. I have indeed seen several injuries from injection of htdraulic oil and hydro test water. Also air paint gun injection. All had an unhappy side effect, and that was the injection of huge amounts of bacteria with ulgy results. The paint gun injection resulted from the removal of the guard, and when injected cost the man most of his hand.

I normally recommend schedule 80 up to about 3000 psi, and then go to schedule 160. At 10,000 and above I suggest the move to a coned and threaded system. I have worked at pressures up to 30,000psi with the traded and coned system. By the way, The velocity in a pin hole is fierce, but the liquid pressure remains the same. I have heard that the injection machines used for needle-less medical inoculations work at 2800 psi. As many hydraulic systems operate at 3000 psi these days the injunction to guard against pinholes is very valid. Last time I made that warning I was accused of being afraid of hydraulics. Worked every day with them for over 20 years. Afraid? No Respectfull? Yes.

Ptree, Obviously, we agree that care should be taken in the design and construction of hydraulic forging systems. Regarding planning for failure and knowing the risks, we also obviously agree that it should be done.

We may not Agree on how to Plan and Build.

While your cautions are good and correct, your hydraulic posts can be read as “don’t do it”. I’d rather educate on how it CAN be done. Hydraulics is a great tool. A hydraulic forging press is FAR safer than a power hammer of ANY variety. Much more controlled and slower motion results in less flying iron and more time to get out of the way. A hydraulic forging press is not right for all work.

There are ways to be safe with hydraulics and they are not rocket science or costly. The hoses and connections on my forging press are on the other side of the press from me. If a hose, connection or pipe leaks, it will not spray on me. THIS is the correct way to guard against a fluid injection. Another safety risk is not replacing things until they leak. That’s asking for trouble.

While I frequently design for 8000 psi at the request of customers, I do not recommend high pressure for most industrial applications. Anything over 2500 psi is “high pressure” to me. More energy is wasted with higher pressures and there is, as we are discussing, many safety concerns. High pressure is required in aircraft for light weight and small component size. Some mobile equipment also benefits from high pressure hydraulics. There is no real need for high pressure hydraulics for the average forging press. Sure, the component size will be a little bigger, but there will be less noise, less heat, far less maintenance, far more safety and less energy use with lower pressure hydraulics. But we must always have more. We must be up to date. We must have the latest. That is another common misconception in hydraulics. 1500 psi is a good number for a forging press of the size we are talking about in my book.

Want to most effectively guard against pinhole leaks? Use lower pressure and keep the human away from the unlikely leak. And maintain the system! NEVER reuse an o-ring or seal. Keep it clean. Replace components when they show wear. Do you replace your tires when they go flat?

The only safe way to do something is to understand what you are working with. YOUR Education is the only true safety measure you can rely on. Cautions are not adequate. If you rely on others, you will be much more likely to get hurt. If you work with something that can hurt you and don't ask about it or don't understand it, you are proving Darwin correct. If you assume a factor of safety of 6 to 1 for pipe, which is a hydraulic industry minimum standard, schedule 80 above 1/8" pipe size is not thick enough for 3000 psi. Most of the hydraulic work we are talking about is with 1/2" to 3/4' pipe size. Schedule 80 1/2" is 2300 psi for 6 to 1 and 3/4" schedule 80 is 1950 psi. Schedule 160 3/4" is 2780 psi for 6 to 1.

What Conduits were you Using for 10,000 and 30,000 psi?

The absolute "hydraulics leak, period", like all absolutes, is rediculous. Same with "a pin hole leak of hydraulic oil if it impacts the hot metal will make a flame-thrower". The reality is that most of the time, IF a small stream hits the usual home shop (small) piece of hot iron, it will hiss and spit, but the oil will quench the iron. CAN there be vapourisation and flame-thrower like activity, yes. But I know of no occurrences like that. I'm not saying you haven't seen it. I'm saying they are very rare. You fail to mention how rare and that is misleading. Most of us are working with pieces of iron that are much smaller and do not hold enough heat to cause a fire. We are in far more danger of a quench oil fire or burn than a hydraulic oil fire. There can be no flame-thrower like activity unless the oil is above it's flash point and there is sufficient oxygenation of the stream of oil.

None of the oil hydraulic equipment I have built or rebuilt, leaks. Accepting leaks is not necessary. If it leaks, someone selected the wrong components or the quality of the components and/or installation is bad. Or the parts are worn. This is common, but does not need to be accepted. "Hydraulics leak" is incorrect. The biggest culprit (after poor installation or maintenance) leading to "hydraulics leak" is tapered pipe fittings. Pipe fittings have no place in a hydraulic system any more. I use them at home when I am mating to an old system or using salvaged components (which is often, grin). With the right quality, techniques and sealants, pipe fittings can work, but are not recommended. SAE o-ring ports, flare fittings, and o-ring face seal fittings are far better. And frequently no more expensive as we have discussed before.

Fire resistant fluids require a significant step up in education. The downsides are sometimes higher toxicity, certainly higher purchase operating and maintenance costs for the fluid, and usually much higher maintenance costs for the equipment. IF used properly, they are safer around hot metal from a fire standpoint. Which is safer from an injection standpoint? The various fire resistant fluids or oil?

I'll stand by my recommendation to use oil and use it wisely for the average small or home shop. We can disagree on that.

As we're discussing this, I'm trying to think of a safe activity. Laying on the couch definitely kills you. Any activity I can think of, that results in accomplishment, has some risk. I can only think of one activity that is "safe". You get enjoyment, give enjoyment, and can get some good exercise. Even that has been known to kill people. But hey, if you die that way, you might well consider yourself a lucky man. Grin. Or woman.

I used swaglok mostly in the 1/2" size up to about 10,000 psi and as stated coned and threaded tubing above. One brand is Autoclave Engineers. Another in High Pressure Equipment. For the swaglow I used stainless steel tubing, seamless, in 0.095 wall. Used both in 3 shift a day, cyclic hydro test, with a pressure test cycle about every 45 seconds.

I do not recommend against hydraulics use. Far from it. Just as you, I recommend that care and prudence be used. I have worked around industrial equipment since 1978. I have seen hydraulics from the USA, England, Germany, Italy, and Japan. I have seen every fitting style and type. Yes, use new o-rings, but people don't. Don't use tapered pipe thread, but people do. Don't seal pipe thread with tape, but people do. Don't underbuild, but people do. In those years, i worked in R & D labs testing some of the very fittings and systems you mention as leak free.

And again I stand pat, hydraulics leak. Not a question of if, but rather when and where. I have seen very high class maintenance and he things oozed, hoses failed, tubing cracked etc. Aluminium manifolds fatiqued and cracked and leaked polygylcol more toxic than oil? enough, on to other things.

How to Measure your Water Flow Rate Q from an Open Pipe: There are times when it's necessary to estimate the flow rate from a stream constrained to flow in a water jet flowing from an open pipe. This flow measurement method doesn't require us to have precision fluid flow measurement instrumentation, other than a straight edge and plumb bob. Flow metres or weirs would likely be more accurate, but sometimes all you have is a filled pipe with a jet of water streaming out. If that's the case, then try this method.

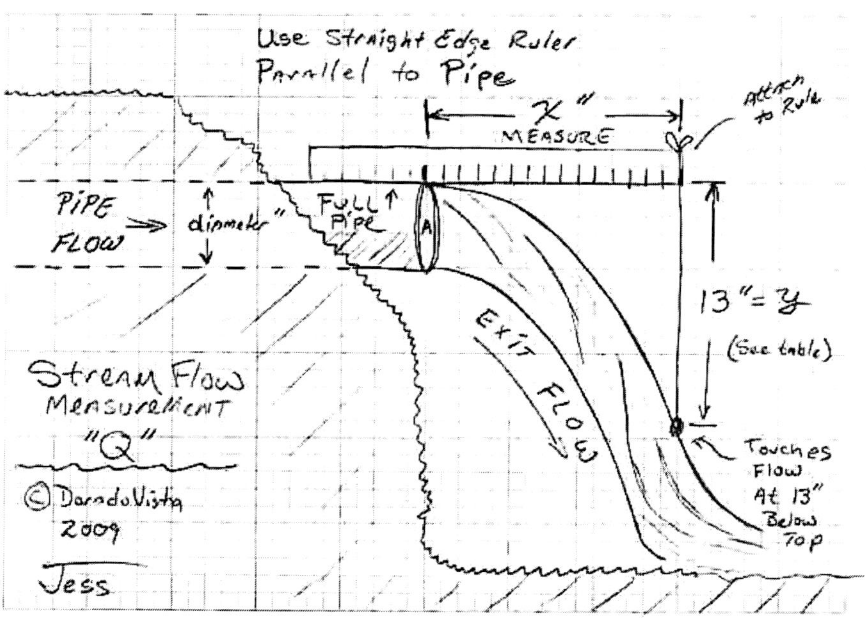

Figure: Pipe Flow Q Direct Measure Diagram A

There are two basic pipe flow measurement methods used to make this estimation. The first pipe flow measurement method involves pipes that are completely filled; that is they have no air in the water pipe above the fluid exiting the pipe, the other method used is a partially filled pipe.

Meanwhile Let's Examine how to Measure the Pipe Flow Rate Q from a Water Jet Squirting from a full Pipe:

What is needed to measure the completely filled pipe flow case is a simple straight edge ruler and a plumb bob marked off at an appropriate pre defined distance, in this case we use 13 inches. To utilise this method one simply measures out an appropriate extra distance parallel to the initial pipe exit flow direction up to an intercept point with the plumb bob line that is 13 inches below that point.

You will move the ruler parallel to the pipeline until the plumb bob just touches the outer edge of the falling stream of w*ater.* By measuring this intercept distance X. for the given 13 inch plumb bob Y. value, you then simply use the table below with your pipe diametre and X value to look up the flow in gallons per minute or GPM. You can convert these values to your desired standard units of measurement as well.

This water flow estimation technique is used in agriculture to measure irrigation flow from an open pipe. This method can also be used for smaller flows, if a stream can be constrained temporarily to completely fill the inside of a pipe and flow out the end above the ground or stream tail water enough to not interfere with the plumb bob. Keep in mind that if the pipe is small enough, direct measurement can be made by filling the bucket and timing the amount of time it takes to fill it, we'll discuss this in another Small Hydro power stream flow measurement post and link to it later.

So, Let's take a look at the Measurement Technique, see Pipe Flow Measurement Procedures, and Diagrams A., B. and C. :

Tools Needed to Measure your Hydropower System Pipe Flow Rate Q:

1. Straight edge ruler marked in inches
2. Plumb bob or thinweight and string at least 2 feet long
3. A pipe completely filled with the water flow jet about 18-24 inchs up in the air above any interference.

Procedure to Measure Potential Hydropower System Flow Q from a Full Flowing Pipe:

1. Tie plumb bob so it dangles 13 inches below the bottom of straight edge ruler
2. Place straight edge ruler along top edge of pipe parallel to exit flow

3. Shift ruler until plumb bob just touches outer surface of the flow at 13 inches below the attachment to ruler.
4. Measure X distance between pipe tip and bob string attachment point.
5. Measure inside diametre of pipe
6. Use 'X' and Diametre with edges of table to read off flow rate in GallonsPer Minute (GPM)

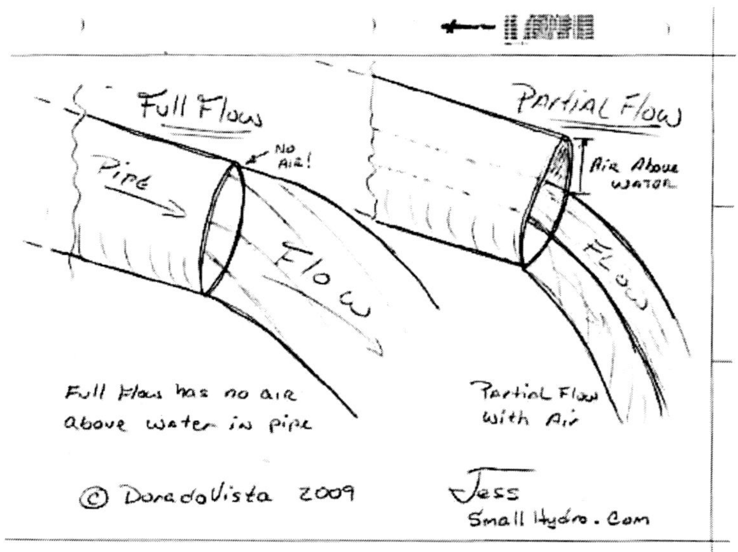

Figure: Pipe Flow Q Direct Measure Diagrams Full & Partial flow D-E

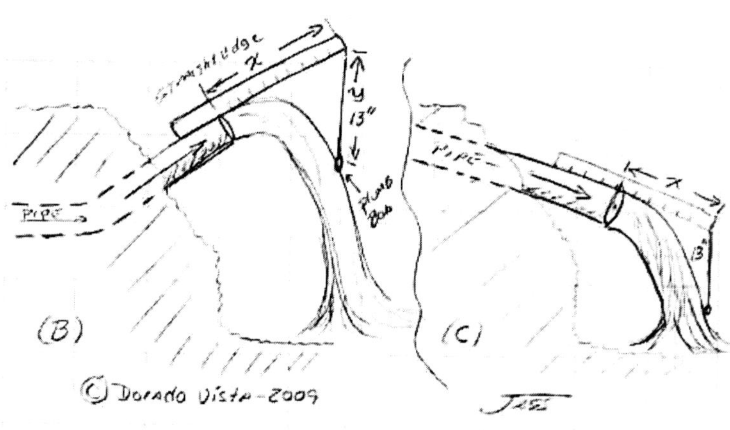

Figure: Pipe Flow Q Direct Measure Diagrams B & C

Some more views of the same measurement, it works the same regardless of pipe angle. Of course bigger errors can happen at steeper angles.

$$GPM = \frac{3.61 \times A \times X}{\sqrt{Y}}$$

Equation for Full Pipe Flow Rate in GPM

Bernoulli's Principle

In fluid dynamics, Bernoulli's principle states that for an inviscid flow, an increase in the speed of the fluid occurs simultaneously with a decrease in pressure or a decrease in the fluid's potential energy. Bernoulli's principle is named after the Dutch-Swiss mathematician Daniel Bernoulli who published his principle in his book *Hydrodynamica* in 1738.

Bernoulli's principle can be applied to various types of fluid flow, resulting in what is loosely denoted as Bernoulli's equation. In fact, there are different forms of the Bernoulli equation for different types of flow. The simple form of Bernoulli's principle is valid for incompressible flows (e.g. most liquid flows) and also for compressible flows (e.g. gases) moving at low Mach numbers. More advanced forms may in some cases be applied to compressible flows at higher Mach numbers.

Bernoulli's principle can be derived from the principle of conservation of energy. This states that, in a steady flow, the sum of all forms of mechanical energy in a fluid along a streamline is the same at all points on that streamline. This requires that the sum of kinetic energy and potential energy remain constant. Thus an increase in the speed of the fluid occurs proportionately with an increase in both its dynamic pressure and kinetic energy, and a decrease in its static pressure and potential energy. If the fluid is flowing out of a reservoir, the sum of all forms of energy is the same on all streamlines because in a reservoir the energy per unit volume (the sum of pressure and gravitational potential $\rho g h$) is the same everywhere.

Bernoulli's principle can also be derived directly from Newton's 2nd law. If a small volume of fluid is flowing horizontally from a region of high pressure to a region of low pressure, then there is more pressure behind than in front. This gives a net force on the volume, accelerating it along the streamline. Fluid particles are subject only to pressure and their own weight. If a fluid is flowing horizontally and along a section of a streamline, where the speed increases it can only be because the fluid on that section has moved from a region of higher pressure to a region of lower pressure; and if its speed decreases, it can only be because it has moved from a region of lower pressure to a region of higher pressure. Consequently, within a fluid flowing horizontally, the highest speed occurs where the pressure is lowest, and the lowest speed occurs where the pressure is highest.

Incompressible Flow Equation

In most flows of liquids, and of gases at low Mach number, the mass density of a fluid parcel can be considered to be constant, regardless of pressure variations in the flow. For this reason the fluid in such flows can be considered to be incompressible and these flows can be described as incompressible flow. Bernoulli performed his experiments on liquids and his equation in its original form is valid only for incompressible flow. A common form of Bernoulli's equation, valid at any arbitrary point along a streamline where gravity is constant, is:

$$\frac{v^2}{2} + gz + \frac{p}{\rho} = \text{constant}$$

where:

$v$is the fluid flow speed at a point on a streamline,

$g$is the acceleration due to gravity,

$z$is the elevation of the point above a reference plane, with the positive z-direction pointing upward – so in the direction opposite to the gravitational acceleration,

$p$is the pressure at the chosen point, and

$\rho$is the density of the fluid at all points in the fluid.

For conservative force fields, Bernoulli's equation can be generalised as:

$$\frac{v^2}{2} + \Psi + \frac{p}{\rho} = \text{constant}$$

where ψ is the force potential at the point considered on the streamline. *E.g.* for the Earth's gravity $\psi = gz$.

The following two assumptions must be met for this Bernoulli equation to apply:

- the flow must be incompressible – even though pressure varies, the density must remain constant along a streamline;
- friction by viscous forces has to be negligible.

By multiplying with the fluid density ρ, equation (A) can be rewritten as:

$$\tfrac{1}{2}\rho v^2+\rho g z+p=\text{constant}$$

or:

$$q+\rho g h=p_0+\rho g z=\text{constant}$$

where:

$q=\tfrac{1}{2}\rho v^2$ is dynamic pressure,

$h=z+\dfrac{p}{\rho g}$ is the piezometric head or hydraulic head (the sum of the elevation z and the pressure head) and

$p_0=p+q$ is the total pressure (the sum of the static pressure p and dynamic pressure q).

The constant in the Bernoulli equation can be normalised. A common approach is in terms of total head or energy head H:

$$H=z+\frac{p}{\rho g}+\frac{v^2}{2g}=h+\frac{v^2}{2g},$$

The above equations suggest there is a flow speed at which pressure is zero, and at even higher speeds the pressure is negative. Most often, gases and liquids are not capable of negative absolute pressure, or even zero pressure, so clearly Bernoulli's equation ceases to be valid before zero pressure is reached. In liquids – when the pressure becomes too low – cavitation occurs. The above equations use a linear relationship between flow speed squared and pressure. At higher flow speeds in gases, or for sound waves in liquid, the changes in mass density become significant so that the assumption of constant density is invalid.

Simplified Form

In many applications of Bernoulli's equation, the change in the $\rho g z$ term along the streamline is so small compared with the other terms it can be ignored. For example, in the case of aircraft in flight, the change in height z along a streamline is so small the $\rho g z$ term can be omitted. This allows the above equation to be presented in the following simplified form:

$$p+q=p_0$$

where p_0 is called total pressure, and q is dynamic pressure. Many authors refer to the pressure p as static pressure to distinguish it from total pressure p_0 and dynamic pressure q. In *Aerodynamics*, L.J. Clancy writes: "To distinguish it from the total and dynamic pressures, the actual pressure of the fluid, which is associated not with its motion but with its state, is often referred to as the static pressure, but where the term pressure alone is used it refers to this static pressure."

The simplified form of Bernoulli's equation can be summarised in the following memorable word equation:

static pressure + dynamic pressure = total pressure

Every point in a steadily flowing fluid, regardless of the fluid speed at that point, has its own unique static pressure p and dynamic pressure q. Their sum $p + q$ is defined to be the total pressure p_0. The significance of Bernoulli's principle can now be summarised as *total pressure is constant along a streamline.*

If the fluid flow is irrotational, the total pressure on every streamline is the same and Bernoulli's principle can be summarised as *total pressure is constant everywhere in the fluid flow.* It is reasonable to assume that irrotational flow exists in any situation where a large body of fluid is flowing past a solid body. Examples are aircraft in flight, and ships moving in open bodies of water. However, it is important to remember that Bernoulli's principle does not apply in the boundary layer or in fluid flow through long pipes.

If the fluid flow at some point along a stream line is brought to rest, this point is called a stagnation point, and at this point the total pressure is equal to the stagnation pressure.

Applicability of Incompressible Flow Equation to Flow of Gases

Bernoulli's equation is sometimes valid for the flow of gases: provided that there is no transfer of kinetic or potential energy from the gas flow to the compression or expansion of the gas. If both the gas pressure and volume change simultaneously, then work will be done on or by the gas.

In this case, Bernoulli's equation – in its incompressible flow form – can not be assumed to be valid. However if the gas process is entirely isobaric, or isochoric, then no work is done on or by the gas, (so the simple energy balance is not upset). According to the gas law, an isobaric or isochoric process is ordinarily the only way to ensure constant density in a gas. Also the gas density will be proportional to the ratio of pressure and absolute temperature, however this ratio will vary upon compression or expansion, no matter what non-zero quantity of heat is added or removed. The only exception is if the net heat transfer is zero, as in a complete thermodynamic cycle, or in an individual is entropic (frictionless adiabatic) process, and even then this reversible process must be reversed, to restore the gas to the original pressure and specific volume, and thus density. Only then is the original, unmodified Bernoulli equation applicable.

In this case the equation can be used if the flow speed of the gas is sufficiently below the speed of sound, such that the variation in density of the gas (due to this effect) along each streamline can be ignored. Adiabatic flow at less than Mach 0.3 is generally considered to be slow enough.

Unsteady Potential Flow

The Bernoulli equation for unsteady potential flow is used in the theory of ocean surface waves and acoustics.

For an irrotational flow, the flow velocity can be described as the gradient $\nabla\varphi$ of a velocity potential φ. In that case, and for a constant density ρ, the momentum equations of the Euler equations can be integrated to:

$$\frac{\partial\varphi}{\partial t}+\tfrac{1}{2}v^2+\frac{p}{\rho}+gz=f(t),$$

which is a Bernoulli equation valid also for unsteady – or time dependent – flows. Here $\partial\varphi/\partial t$ denotes the partial derivative of the velocity potential φ with respect to time t, and $v = |\nabla\varphi|$ is the flow speed. The function $f(t)$ depends only on time and not on position in the fluid. As a result, the Bernoulli equation at some moment t does not only apply along a certain streamline, but in the whole fluid domain. This is also true for the special case of a steady irrotational flow, in which case f is a constant.

Further $f(t)$ can be made equal to zero by incorporating it into the velocity potential using the transformation

$$\Phi=\varphi-\int_{t_0}^{t} f(\tau)\,\mathrm{d}\tau, \text{ resulting in } \frac{\partial\Phi}{\partial t}+\tfrac{1}{2}v^2+\frac{p}{\rho}+gz=0.$$

Note that the relation of the potential to the flow velocity is unaffected by this transformation:

$$\nabla\Phi=\nabla\varphi.$$

The Bernoulli equation for unsteady potential flow also appears to play a central role in Luke's variational principle, a variational description of free-surface flows using the Lagrangian (not to be confused with Lagrangian coordinates).

Compressible Flow Equation

Bernoulli developed his principle from his observations on liquids, and his equation is applicable only to incompressible fluids, and compressible fluids up to approximately Mach number 0.3. It is possible to use the fundamental principles of physics to develop similar equations applicable to compressible fluids.

There are numerous equations, each tailored for a particular application, but all are analogous to Bernoulli's equation and all rely on nothing more than the fundamental principles of physics such as Newton's laws of motion or the first law of thermodynamics.

Compressible Flow in Fluid Dynamics

For a compressible fluid, with a barotropic equation of state, and under the action of conservative forces,

$$\frac{v^2}{2} + \int_{p_1}^{p} \frac{d\tilde{p}}{\rho(\tilde{p})} + \Psi = \text{constant}$$

where:

p is the pressure

ρ is the density

v is the flow speed

ψ is the potential associated with the conservative force field, often the gravitational potential.

In engineering situations, elevations are generally small compared to the size of the Earth, and the time scales of fluid flow are small enough to consider the equation of state as adiabatic.

In this case, the above equation becomes

$$\frac{v^2}{2} + gz + \left(\frac{\gamma}{\gamma - 1}\right)\frac{p}{\rho} = \text{constant}$$

where, in addition to the terms listed above:

γ is the ratio of the specific heats of the fluid

g is the acceleration due to gravity

z is the elevation of the point above a reference plane

In many applications of compressible flow, changes in elevation are negligible compared to the other terms, so the term gz can be omitted.

A very useful form of the equation is then:

$$\frac{v^2}{2}+\left(\frac{\gamma}{\gamma-1}\right)\frac{p}{\rho}=\left(\frac{\gamma}{\gamma-1}\right)\frac{p_0}{\rho_0}$$

where:

p_0 is the total pressure

ρ_0 is the total density

Compressible Flow in Thermodynamics

Another useful form of the equation, suitable for use in thermodynamics, is:

$$\frac{v^2}{2}+\Psi+w=\text{constant.}$$

Here w is the enthalpy per unit mass, which is also often written as h (not to be confused with "head" or "height").

Note that $w=\epsilon+\frac{p}{\rho}$ where ε is the thermodynamic energy per unit mass, also known as the specific internal energy.

The constant on the right hand side is often called the Bernoulli constant and denoted b. For steady inviscid adiabatic flow with no additional sources or sinks of energy, b is constant along any given streamline. More generally, when b may vary along streamlines, it still proves a useful parameter, related to the "head" of the fluid.

When the change in ψ can be ignored, a very useful form of this equation is:

$$\frac{v^2}{2}+w=w_0$$

where w_0 is total enthalpy. For a calorically perfect gas such as an ideal gas, the enthalpy is directly proportional to the temperature, and this leads to the concept of the total (or stagnation) temperature.

When shock waves are present, in a reference frame in which the shock is stationary and the flow is steady, many of the parameters in the Bernoulli equation suffer abrupt changes in passing through the shock.

The Bernoulli parameter itself, however, remains unaffected. An exception to this rule is radiative shocks, which violate the assumptions leading to the Bernoulli equation, namely the lack of additional sinks or sources of energy.

Bernoulli equation for incompressible fluids The Bernoulli equation for incompressible fluids can be derived by integrating the Euler equations, or applying the law of conservation of energy in two sections along a streamline, ignoring viscosity, compressibility, and thermal effects.

The simplest derivation is to first ignore gravity and consider constrictions and expansions in pipes that are otherwise straight, as seen in Venturi effect. Let the x axis be directed down the axis of the pipe. Define a parcel of fluid moving through a pipe with cross-sectional area "A", the length of the parcel is "dx", and the volume of the parcel A dx. If mass density is ρ, the mass of the parcel is density multiplied by its volume $m = \rho A$ dx. The change in pressure over distance dx is "dp" and flow velocity $v = dx / dt$.

Apply Newton's Second Law of Motion Force F =mass acceleration and recognising that the effective force on the parcel of fluid is -A dp. If the pressure decreases along the length of the pipe, dp is negative but the force resulting in flow is positive along the x axis.

$$m\frac{\mathrm{d}v}{\mathrm{d}t} = F$$

$$\rho A \,\mathrm{d}x \frac{\mathrm{d}v}{\mathrm{d}t} = -A\,\mathrm{d}p$$

$$\rho\frac{\mathrm{d}v}{\mathrm{d}t} = -\frac{\mathrm{d}p}{\mathrm{d}x}$$

In steady flow the velocity is constant with respect to time, $v = v(x) = v(x(t))$, so v itself is not directly a function of time t. It is only when the parcel moves through x that the cross sectional area changes: v depends on t only through the cross-sectional position $x(t)$.

$$\frac{\mathrm{d}v}{\mathrm{d}t} = \frac{\mathrm{d}v}{\mathrm{d}x}\frac{\mathrm{d}x}{\mathrm{d}t} = \frac{\mathrm{d}v}{\mathrm{d}x}v = \frac{d}{\mathrm{d}x}\left(\frac{v^2}{2}\right).$$

With density ρ constant, the equation of motion can be written as

$$\frac{\mathrm{d}}{\mathrm{d}x}\left(\rho\frac{v^2}{2} + p\right) = 0$$

by integrating with respect to x

$$\frac{v^2}{2} + \frac{p}{\rho} = C$$

where C is a constant, sometimes referred to as the Bernoulli constant. It is not a universal constant, but rather a constant of a particular fluid system. The deduction is: where the speed is large, pressure is low and vice versa. In the above derivation, no external work-energy principle is invoked. Rather, Bernoulli's principle was inherently derived by a simple manipulation of the momentum equation.

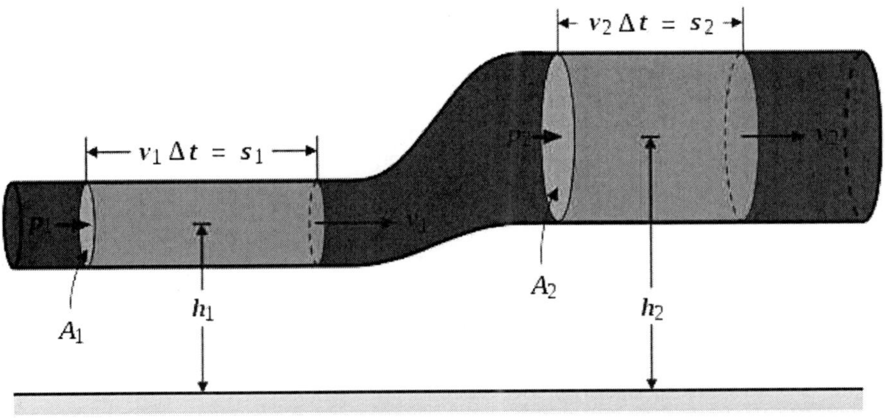

Figure: A streamtube of fluid moving to the right. Indicated are pressure, elevation, flow speed, distance (s), and cross-sectional area.

Another way to derive Bernoulli's principle for an incompressible flow is by applying conservation of energy. In the form of the work-energy theorem, stating that

the change in the kinetic energy E_{kin} *of the system equals the net work* W *done on the system*;

$$W = \Delta E_{\text{kin}}.$$

Therefore,

the work done by the forces in the fluid = increase in kinetic energy.

The system consists of the volume of fluid, initially between the cross-sections A_1 and A_2. In the time interval Δt fluid elements initially at the inflow cross-section A_1 move over a distance $s_1 = v_1 \Delta t$, while at the outflow cross-section the fluid moves away from cross-section A_2 over a distance $s_2 = v_2 \Delta t$.

The displaced fluid volumes at the inflow and outflow are respectively $A_1 s_1$ and $A_2 s_2$. The associated displaced fluid masses are – when ρ is the fluid's mass density – equal to density times volume, so $\rho A_1 s_1$ and $\rho A_2 s_2$. By mass conservation, these two masses displaced in the time interval Δt have to be equal, and this displaced mass is denoted by Δm:

$$\rho A_1 s_1 = \rho A_1 v_1 \Delta t = \Delta m,$$
$$\rho A_2 s_2 = \rho A_2 v_2 \Delta t = \Delta m.$$

The work done by the forces consists of two parts:

- The *work done by the pressure* acting on the areas A_1 and A_2

$$W_{\text{pressure}} = F_{1,\text{pressure}}\, s_1 - F_{2,\text{pressure}}\, s_2 = p_1 A_1 s_1 - p_2 A_2 s_2 = \Delta m \frac{p_1}{\rho} - \Delta m \frac{p_2}{\rho}.$$

- The *work done by gravity*: the gravitational potential energy in the volume $A_1\, s_1$ is lost, and at the outflow in the volume $A_2\, s_2$ is gained. So, the change in gravitational potential energy $\Delta E_{\text{pot,gravity}}$ in the time interval Δt is

$$\Delta E_{\text{pot,gravity}} = \Delta m\, g z_2 - \Delta m\, g z_1.$$

Now, the work by the force of gravity is opposite to the change in potential energy, $W_{\text{gravity}} = -\Delta E_{\text{pot,gravity}}$: while the force of gravity is in the negative z-direction, the work—gravity force times change in elevation—will be negative for a positive elevation change $\Delta z = z_2 - z_1$, while the corresponding potential energy change is positive. So:

$$W_{\text{gravity}} = -\Delta E_{\text{pot,gravity}} = \Delta m\, g z_1 - \Delta m\, g z_2.$$

And the total work done in this time interval Δt is

$$W = W_{\text{pressure}} + W_{\text{gravity}}.$$

The *increase in kinetic energy* is

$$\Delta E_{\text{kin}} = \frac{1}{2}\Delta m v_2^2 - \frac{1}{2}\Delta m v_1^2.$$

Putting these together, the work-kinetic energy theorem $W = \Delta E_{\text{kin}}$ gives:

$$\Delta m \frac{p_1}{\rho} - \Delta m \frac{p_2}{\rho} + \Delta m\, g z_1 - \Delta m\, g z_2 = \frac{1}{2}\Delta m v_2^2 - \frac{1}{2}\Delta m v_1^2$$

or

$$\frac{1}{2}\Delta m v_1^2 + \Delta m\, g z_1 + \Delta m \frac{p_1}{\rho} = \frac{1}{2}\Delta m v_2^2 + \Delta m\, g z_2 + \Delta m \frac{p_2}{\rho}.$$

After dividing by the mass $\Delta m = \rho A_1\, v_1\, \Delta t = \rho A_2\, v_2\, \Delta t$ the result is:

$$\frac{1}{2}v_1^2 + gz_1 + \frac{p_1}{\rho} = \frac{1}{2}v_2^2 + gz_2 + \frac{p_2}{\rho}$$

or, as stated in the first paragraph:

$$\frac{v^2}{2} + gz + \frac{p}{\rho} = C$$ (Eqn. 1), Which is also Equation (A)

Further division by g produces the following equation. Note that each term can be described in the length dimension (such as metres). This is the head equation derived from Bernoulli's principle:

$$\frac{v^2}{2g} + z + \frac{p}{\rho g} = C$$ (Eqn. 2a)

The middle term, z, represents the potential energy of the fluid due to its elevation with respect to a reference plane. Now, z is called the elevation head and given the designation $z_{elevation}$.

A free falling mass from an elevation $z > 0$ (in a vacuum) will reach a speed $v = \sqrt{2gz}$,

when arriving at elevation $z = 0$. Or when we rearrange it as a *head*:

$$h_v = \frac{v^2}{2g}$$

The term $v^2 / (2g)$ is called the *velocity head*, expressed as a length measurement. It represents the internal energy of the fluid due to its motion.

The hydrostatic pressure p is defined as

$p = p_0 - \rho gz$, with p_0 some reference pressure, or when we rearrange it as a *head*: $\psi = \frac{p}{\rho g}$

The term $p / (\rho g)$ is also called the *pressure head*, expressed as a length measurement. It represents the internal energy of the fluid due to the pressure exerted on the container.

When we combine the head due to the flow speed and the head due to static pressure with the elevation above a reference plane, we obtain a simple relationship useful for incompressible fluids using the velocity head, elevation head, and pressure head.

$$h_v + z_{elevation} + \psi = C$$ (Eqn. 2b)

If we were to multiply Eqn. 1 by the density of the fluid, we would get an equation with three pressure terms:

$$\frac{\rho v^2}{2} + \rho g z + p = C \quad \text{(Eqn. 3)}$$

We note that the pressure of the system is constant in this form of the Bernoulli Equation. If the static pressure of the system (the far right term) increases, and if the pressure due to elevation (the middle term) is constant, then we know that the dynamic pressure (the left term) must have decreased. In other words, if the speed of a fluid decreases and it is not due to an elevation difference, we know it must be due to an increase in the static pressure that is resisting the flow. All three equations are merely simplified versions of an energy balance on a system.

Bernoulli equation for compressible fluids The derivation for compressible fluids is similar. Again, the derivation depends upon (1) conservation of mass, and (2) conservation of energy. The interval of time Δt, the amount of mass passing through the boundary defined by the area A_1 is equal to the amount of mass passing outwards through the boundary defined by the area A_2:

$$0 = \Delta M_1 - \Delta M_2 = \rho_1 A_1 v_1 \Delta t - \rho_2 A_2 v_2 \Delta t \,.$$

Conservation of energy is applied in a similar manner: It is assumed that the change in energy of the volume of the streamtube bounded by A_1 and A_2 is due entirely to energy entering or leaving through one or the other of these two boundaries. Clearly, in a more complicated situation such as a fluid flow coupled with radiation, such conditions are not met. Nevertheless, assuming this to be the case and assuming the flow is steady so that the net change in the energy is zero,

$$0 = \Delta E_1 - \Delta E_2$$

where ΔE_1 and ΔE_2 are the energy entering through A_1 and leaving through A_2, respectively.

The energy entering through A_1 is the sum of the kinetic energy entering, the energy entering in the form of potential gravitational energy of the fluid, the fluid thermodynamic energy entering, and the energy entering in the form of mechanical $p\,dV$ work:

$$\Delta E_1 = \left[\frac{1}{2}\rho_1 v_1^2 + \Psi_1 \rho_1 + \epsilon_1 \rho_1 + p_1\right] A_1 v_1 \Delta t$$

where $\psi = gz$ is a force potential due to the Earth's gravity, g is acceleration due to gravity, and z is elevation above a reference plane.

A similar expression for ΔE_2 may easily be constructed. So now setting $0 = \Delta E_1 - \Delta E_2$:

$$0 = \left[\frac{1}{2}\rho_1 v_1^2 + \Psi_1\rho_1 + \epsilon_1\rho_1 + p_1\right]A_1 v_1 \Delta t - \left[\frac{1}{2}\rho_2 v_2^2 + \Psi_2\rho_2 + \epsilon_2\rho_2 + p_2\right]A_2 v_2 \Delta t$$

which can be rewritten as:

$$0 = \left[\frac{1}{2}v_1^2 + \Psi_1 + \epsilon_1 + \frac{p_1}{\rho_1}\right]\rho_1 A_1 v_1 \Delta t - \left[\frac{1}{2}v_2^2 + \Psi_2 + \epsilon_2 + \frac{p_2}{\rho_2}\right]\rho_2 A_2 v_2 \Delta t$$

Now, using the previously-obtained result from conservation of mass, this may be simplified to obtain

$$\frac{1}{2}v^2 + \Psi + \epsilon + \frac{p}{\rho} = \text{constant} \equiv b$$

which is the Bernoulli equation for compressible flow.

Real-world Application

In modern everyday life there are many observations that can be successfully explained by application of Bernoulli's principle, even though no real fluid is entirely inviscid and a small viscosity often has a large effect on the flow.

Bernoulli's principle can be used to calculate the lift force on an airfoil if the behaviour of the fluid flow in the vicinity of the foil is known. For example, if the air flowing past the top surface of an aircraft wing is moving faster than the air flowing past the bottom surface, then Bernoulli's principle implies that the pressure on the surfaces of the wing will be lower above than below. This pressure difference results in an upwards lift force. Whenever the distribution of speed past the top and bottom surfaces of a wing is known, the lift forces can be calculated (to a good approximation) using Bernoulli's equations – established by Bernoulli over a century before the first man-made wings were used for the purpose of flight. Bernoulli's principle does not explain why the air flows faster past the top of the wing and slower past the underside. To understand why, it is helpful to understand circulation, the Kutta condition, and the Kutta–Joukowski theorem.

The carburettor used in many reciprocating engines contains a venturi to create a region of low pressure to draw fuel into the carburettor and mix it thoroughly with the incoming air. The low pressure in the throat of a venturi can be explained by Bernoulli's principle; in the narrow throat, the air is moving at its fastest speed and therefore it is at its lowest pressure.

The Pitot tube and static port on an aircraft are used to determine the airspeed of the aircraft. These two devices are connected to the airspeed indicator which determines the dynamic pressure of the airflow past the aircraft. Dynamic pressure is the difference between stagnation pressure and static pressure. Bernoulli's principle is used to calibrate the airspeed indicator so that it displays the indicated airspeed appropriate to the dynamic pressure.

The flow speed of a fluid can be measured using a device such as a Venturi metre or an orifice plate, which can be placed into a pipeline to reduce the diametre of the flow. For a horizontal device, the continuity equation shows that for an incompressible fluid, the reduction in diametre will cause an increase in the fluid flow speed. Subsequently Bernoulli's principle then shows that there must be a decrease in the pressure in the reduced diametre region. This phenomenon is known as the Venturi effect. The maximum possible drain rate for a tank with a hole or tap at the base can be calculated directly from Bernoulli's equation, and is found to be proportional to the square root of the height of the fluid in the tank. This is Torricelli's law, showing that Torricelli's law is compatible with Bernoulli's principle. Viscosity lowers this drain rate. This is reflected in the discharge coefficient, which is a function of the Reynolds number and the shape of the orifice. In open-channel hydraulics, a detailed analysis of the Bernoulli theorem and its extension were recently (2009) developed. It was proved that the depth-averaged specific energy reaches a minimum in converging accelerating free-surface flow over weirs and flumes (also). Further, in general, a channel control with minimum specific energy in curvilinear flow is not isolated from water waves, as customary state in open-channel hydraulics.

The Bernoulli grip relies on this principle to create a non-contact adhesive force between a surface and the gripper.

Misunderstandings About the Generation of Lift

Many explanations for the generation of lift (on airfoils, propeller blades, etc.) can be found; but some of these explanations can be misleading, and some are false. This has been a source of heated discussion over the years. In particular, there has been debate about whether lift is best explained by Bernoulli's principle or Newton's laws of motion. Modern writings agree that both Bernoulli's principle and Newton's laws are relevant and either can be used to correctly describe lift.

Several of these explanations use the Bernoulli principle to connect the flow kinematics to the flow-induced pressures. In cases of incorrect (or partially correct) explanations relying on the Bernoulli principle, the errors generally occur in the assumptions on the flow kinematics and how these are produced. It is not the Bernoulli principle itself that is questioned because this principle is well established.

Terminology in Fluid Dynamics

The concept of pressure is central to the study of both fluid statics and fluid dynamics. A pressure can be identified for every point in a body of fluid, regardless of whether the fluid is in motion or not. Pressure can be measured using an aneroid, Bourdon tube, mercury column, or various other methods.

Some of the terminology that is necessary in the study of fluid dynamics is not found in other similar areas of study. In particular, some of the terminology used in fluid dynamics is not used in fluid statics.

Terminology in Incompressible Fluid Dynamics

The concepts of total pressure and dynamic pressure arise from Bernoulli's equation and are significant in the study of all fluid flows. (These two pressures are not pressures in the usual sense—they cannot be measured using an aneroid, Bourdon tube or mercury column.) To avoid potential ambiguity when referring to pressure in fluid dynamics, many authors use the term static pressure to distinguish it from total pressure and dynamic pressure. Static pressure is identical to pressure and can be identified for every point in a fluid flow field.

In Aerodynamics, L.J. Clancy Writes: To distinguish it from the total and dynamic pressures, the actual pressure of the fluid, which is associated not with its motion but with its state, is often referred to as the static pressure, but where the term pressure alone is used it refers to this static pressure.

A point in a fluid flow where the flow has come to rest (i.e. speed is equal to zero adjacent to some solid body immersed in the fluid flow) is of special significance. It is of such importance that it is given a special name—a stagnation point. The static pressure at the stagnation point is of special significance and is given its own name—stagnation pressure. In incompressible flows, the stagnation pressure at a stagnation point is equal to the total pressure throughout the flow field.

Terminology in Compressible Fluid Dynamics

In a compressible fluid, such as air, the temperature and density are essential when determining the state of the fluid. In addition to the concept of total pressure (also known as stagnation pressure), the concepts of total (or stagnation) temperature and total (or stagnation) density are also essential in any study of compressible fluid flows. To avoid potential ambiguity when referring to temperature and density, many authors use the terms static temperature and static density. Static temperature is identical to temperature; and static density is identical to density; and both can be identified for every point in a fluid flow field.

The temperature and density at a stagnation point are called stagnation temperature and stagnation density.

A similar approach is also taken with the thermodynamic properties of compressible fluids. Many authors use the terms total (or stagnation) enthalpy and total (or stagnation) entropy. The terms static enthalpy and static entropy appear to be less common, but where they are used they mean nothing more than enthalpy and entropy respectively, and the prefix "static" is being used to avoid ambiguity with their 'total' or 'stagnation' counterparts. Because the 'total' flow conditions are defined by is entropically bringing the fluid to rest, the total (or stagnation) entropy is by definition always equal to the "static" entropy.

Euler Equations (Fluid Dynamics)

In fluid dynamics, the Euler equations are a set of equations governing inviscid flow. They are named after Leonhard Euler. The equations represent conservation of mass (continuity), momentum, and energy, corresponding to the Navier–Stokes equations with zero viscosity and heat conduction terms. Historically, only the continuity and momentum equations have been derived by Euler. However, fluid dynamics literature often refers to the full set – including the energy equation – together as "the Euler equations".

Like the Navier-Stokes equations, the Euler equations are usually written in one of two forms: the "conservation form" and the "non-conservation form". The conservation form emphasizes the physical interpretation of the equations as conservation laws through a control volume fixed in space. The non-conservation form emphasises changes to the state of a control volume as it moves with the fluid.

The Euler equations can be applied to compressible as well as to incompressible flow – using either an appropriate equation of state or assuming that the divergence of the flow velocity field is zero, respectively.

History

The Euler equations first appeared in published form in Euler's article "Principes généraux du mouvement des fluides," published in Mémoires de l'Academie des Sciences de Berlin in 1757. They were among the first partial differential equations to be written down. At the time Euler published his work, the system of equations consisted of the momentum and continuity equations, thus it was underdetermined except in the case of an incompressible fluid. An additional equation, which was later to be called the adiabatic condition, was supplied by Pierre-Simon Laplace in 1816.

During the second half of the 19th century, it was found that the equation related to the conservation of energy must at all times be kept, while the adiabatic condition is a consequence of the fundamental laws in the case of smooth solutions. With the discovery of the special theory of relativity, the concepts of energy density, momentum density, and stress were unified into the concept of the stress-energy tensor, and energy and momentum were likewise unified into a single concept, the energy-momentum vector.

Conservation and Component Form

In differential form, the equations are:

$$\frac{\partial \rho}{\partial t} + \nabla\cdot(\rho \mathbf{u}) = 0$$

$$[1.2ex] \quad \frac{\partial \rho \mathbf{u}}{\partial t} + \nabla\cdot(\mathbf{u} \otimes (\rho \mathbf{u})) + \nabla p = \mathbf{0}$$

$$[1.2ex] \quad \frac{\partial E}{\partial t} + \nabla\cdot(\mathbf{u}(E + p)) = 0,$$

where

- ρ is the fluid mass density,
- u is the fluid velocity vector, with components u, v, and w,
- $E = \rho e + ½ \rho (u^2 + v^2 + w^2)$ is the total energy per unit volume, with e being the internal energy per unit mass for the fluid,

- p is the pressure,
- $\otimes$ denotes the tensor product, and
- being the zero vector.

These equations may be expressed in subscript notation. The second equation includes the divergence of a dyadic product, and may be clearer in subscript notation:

$$\frac{\partial \rho}{\partial t} + \sum_{i=1}^{3} \frac{\partial(\rho u_i)}{\partial x_i} = 0,$$

$$\frac{\partial(\rho u_j)}{\partial t} + \sum_{i=1}^{3} \frac{\partial(\rho u_i u_j)}{\partial x_i} + \frac{\partial p}{\partial x_j} = 0,$$

$$\frac{\partial E}{\partial t} + \sum_{i=1}^{3} \frac{\partial((E+p)u_i)}{\partial x_i} = 0,$$

where the i and j subscripts label the three Cartesian components: $(x_1, x_2, x_3) = (x, y, z)$ and $(u_1, u_2, u_3) = (u, v, w)$. These equations may be more succinctly expressed using abstract index notation in which matched indices imply a sum over those indices and $\partial_t = \frac{\partial}{\partial t}$ and $\partial_i = \frac{\partial}{\partial x_i}$:

$$\partial_t \rho + \partial_i(\rho u_i) = 0$$

$$\partial_t(\rho u_j) + \partial_i(\rho u_i u_j) + \partial_j p = 0$$

$$\partial_t E + \partial_i((E+p)u_i) = 0$$

Note that the above equations are expressed in conservation form, as this format emphasizes their physical origins (and is often the most convenient form for computational fluid dynamics simulations). By subtracting the velocity times the mass conservation term, the second equation (momentum conservation), can also be expressed as:

$$[\partial_t(\rho u_j) + \partial_i(\rho u_i u_j) + \partial_j p] - u_j[\partial_t \rho + \partial_i(\rho u_i)] = \rho \partial_t u_j + \rho u_i \partial_i u_j + \partial_j p = 0$$

or, in vector notation:

$$\rho\left(\frac{\partial}{\partial t} + \mathbf{u}\cdot\nabla\right)\mathbf{u} + \nabla p = \mathbf{0}$$

but this form for the momentum conservation equation obscures the direct connection between the Euler equations and Newton's second law of motion. Similarly, by subtracting the velocity times the above momentum conservation term, the third equation (energy conservation), can also be expressed as:

$$\partial_t e + \partial_i (e u_i) + p \partial_i u_i = 0$$

or

$$\frac{\partial e}{\partial t} + \nabla \cdot (e\mathbf{u}) + p\nabla \cdot \mathbf{u} = 0$$

Conservation and Vector Form

In vector and conservation form, the Euler equations become:

$$\frac{\partial \mathbf{m}}{\partial t} + \frac{\partial \mathbf{f}_x}{\partial x} + \frac{\partial \mathbf{f}_y}{\partial y} + \frac{\partial \mathbf{f}_z}{\partial z} = \mathbf{0},$$

where

$$\mathbf{m} = \begin{pmatrix} \rho \\ \rho u \\ \rho v \\ \rho w \\ E \end{pmatrix};$$

$$\mathbf{f}_x = \begin{pmatrix} \rho u \\ p + \rho u^2 \\ \rho uv \\ \rho uw \\ u(E+p) \end{pmatrix}; \quad \mathbf{f}_y = \begin{pmatrix} \rho v \\ \rho uv \\ p + \rho v^2 \\ \rho vw \\ v(E+p) \end{pmatrix}; \quad \mathbf{f}_z = \begin{pmatrix} \rho w \\ \rho uw \\ \rho vw \\ p + \rho w^2 \\ w(E+p) \end{pmatrix}.$$

This form makes it clear that f_x, f_y and f_z are fluxes.

The equations above thus represent conservation of mass, three components of momentum, and energy. There are thus five equations and six unknowns. Closing the system requires an equation of state; the most commonly used is the ideal gas law (i.e. $p = \rho\,(\gamma\text{-}1)\,e$, where ρ is the density, γ is the adiabatic index, and e the internal energy).

The extra terms involving p may be interpreted as the mechanical work done on a fluid element by its neighbour fluid elements. These terms sum to zero in an incompressible fluid.

The well-known Bernoulli's equation can be derived by integrating Euler's equation along a streamline, under the assumption of constant density and a sufficiently stiff equation of state.

Non-conservation form with Flux Jacobians

Expanding the fluxes can be an important part of constructing numerical solvers, for example by exploiting (approximate) solutions to the Riemann problem. From the original equations as given above in vector and conservation form, the equations are written in a non-conservation form as:

$$\frac{\partial \mathbf{m}}{\partial t} + \mathbf{A}_x \frac{\partial \mathbf{m}}{\partial x} + \mathbf{A}_y \frac{\partial \mathbf{m}}{\partial y} + \mathbf{A}_z \frac{\partial \mathbf{m}}{\partial z} = \mathbf{0}.$$

where A_x, A_y and A_z are called the flux Jacobians, which are matrices equal to:

$$\mathbf{A}_x = \frac{\partial \mathbf{f}_x(\mathbf{s})}{\partial \mathbf{s}}, \qquad \mathbf{A}_y = \frac{\partial \mathbf{f}_y(\mathbf{s})}{\partial \mathbf{s}} \qquad \text{and} \qquad \mathbf{A}_z = \frac{\partial \mathbf{f}_z(\mathbf{s})}{\partial \mathbf{s}}.$$

Here, the flux Jacobians A_x, A_y and A_z are still functions of the state vector m, so this form of the Euler equations is nonlinear, just like the original equations. This non-conservation form is equivalent to the original Euler equations in conservation form, at least in regions where the state vector m varies smoothly.

Flux Jacobians for an Ideal Gas

The ideal gas law is used as the equation of state, to derive the full Jacobians in matrix form, as given below:

The x-direction flux Jacobian:

$$\mathbf{A}_x = \begin{bmatrix} 0 & 1 & 0 & 0 & 0 \\ \hat{\gamma}H - u^2 - a^2 & (3-\gamma)u & -\hat{\gamma}v & -\hat{\gamma}w & \hat{\gamma} \\ -uv & v & u & 0 & 0 \\ -uw & w & 0 & u & 0 \\ u[(\gamma-2)H - a^2] & H - \hat{\gamma}u^2 & -\hat{\gamma}uv & -\hat{\gamma}uw & \gamma u \end{bmatrix}.$$

The y-direction flux Jacobian:

$$\mathbf{A}_y = \begin{bmatrix} 0 & 0 & 1 & 0 & 0 \\ -vu & v & u & 0 & 0 \\ \hat{\gamma}H - v^2 - a^2 & -\hat{\gamma}u & (3-\gamma)v & -\hat{\gamma}w & \hat{\gamma} \\ -vw & 0 & w & v & 0 \\ v[(\gamma-2)H - a^2] & -\hat{\gamma}uv & H - \hat{\gamma}v^2 & -\hat{\gamma}vw & \gamma v \end{bmatrix}.$$

The z-direction flux Jacobian:

$$\mathbf{A}_z = \begin{bmatrix} 0 & 0 & 0 & 1 & 0 \\ -uw & w & 0 & u & 0 \\ -vw & 0 & w & v & 0 \\ \hat{\gamma}H - w^2 - a^2 & -\hat{\gamma}u & -\hat{\gamma}v & (3-\gamma)w & \hat{\gamma} \\ w[(\gamma-2)H - a^2] & -\hat{\gamma}uw & -\hat{\gamma}vw & H - \hat{\gamma}w^2 & \gamma w \end{bmatrix}.$$

Where $\hat{\gamma} = \gamma - 1$.

The total enthalpy H is given by:

$$H = \frac{E}{\rho} + \frac{p}{\rho},$$

and the speed of sound a is given as:

$$a = \sqrt{\frac{\gamma p}{\rho}} = \sqrt{(\gamma-1)\left[H - \frac{1}{2}\left(u^2 + v^2 + w^2\right)\right]}.$$

Linearised Form

The linearised Euler equations are obtained by linearisation of the Euler equations in non-conservation form with flux Jacobians, around a state $m = m_0$, and are given by:

$$\frac{\partial \mathbf{m}}{\partial t} + \mathbf{A}_{x,0}\frac{\partial \mathbf{m}}{\partial x} + \mathbf{A}_{y,0}\frac{\partial \mathbf{m}}{\partial y} + \mathbf{A}_{z,0}\frac{\partial \mathbf{m}}{\partial z} = \mathbf{0},$$

where $A_{x,0}$, $A_{y,0}$ and $A_{z,0}$ are the values of respectively A_x, A_y and A_z at some reference state $m = m_0$.

Uncoupled Wave Equations for the Linearised One-dimensional Case

The Euler equations can be transformed into uncoupled wave equations if they are expressed in characteristic variables instead of conserved variables. As an example, the one-dimensional (1-D) Euler equations in linear flux-Jacobian form is considered:

$$\frac{\partial \mathbf{m}}{\partial t} + \mathbf{A}_{x,0}\frac{\partial \mathbf{m}}{\partial x} = \mathbf{0}.$$

The matrix $A_{x,0}$ is diagonalizable, which means it can be decomposed into:

$$\mathbf{A}_{x,0} = \mathbf{P}\mathbf{\Lambda}\mathbf{P}^{-1},$$

$$\mathbf{P} = [\mathbf{r}_1, \mathbf{r}_2, \mathbf{r}_3] = \begin{bmatrix} 1 & 1 & 1 \\ u-a & u & u+a \\ H-ua & \frac{1}{2}u^2 & H+ua \end{bmatrix},$$

$$\mathbf{\Lambda} = \begin{bmatrix} \lambda_1 & 0 & 0 \\ 0 & \lambda_2 & 0 \\ 0 & 0 & \lambda_3 \end{bmatrix} = \begin{bmatrix} u-a & 0 & 0 \\ 0 & u & 0 \\ 0 & 0 & u+a \end{bmatrix}.$$

Here r_1, r_2, r_3 are the right eigenvectors of the matrix $A_{x,0}$ corresponding with the eigenvalues λ_1, λ_2 and λ_3.

Defining the *characteristic variables* as:

$$\mathbf{w} = \mathbf{P}^{-1}\mathbf{m},$$

Since $A_{x,0}$ is constant, multiplying the original 1-D equation in flux-Jacobian form with P^{-1} yields:

$$\frac{\partial \mathbf{w}}{\partial t} + \mathbf{\Lambda}\frac{\partial \mathbf{w}}{\partial x} = \mathbf{0}$$

The equations have been essentially decoupled and turned into three wave equations, with the eigenvalues being the wave speeds. The variables w_i are called *Riemann invariants* or, for general hyperbolic systems, they are called *characteristic variables.*

Shock Waves

The Euler equations are nonlinear hyperbolic equations and their general solutions are waves. Much like the familiar oceanic waves, waves described by the Euler Equations 'break' and so-called shock waves are formed; this is a nonlinear effect and represents the solution becoming multi-valued. Physically this represents a breakdown of the assumptions that led to the formulation of the differential equations, and to extract further information from the equations we must go back to the more fundamental integral form. Then, weak solutions are formulated by working in 'jumps' (discontinuities) into the flow quantities – density, velocity, pressure, entropy – using the Rankine–Hugoniot shock conditions. Physical quantities are rarely discontinuous; in real flows, these discontinuities are smoothed out by viscosity.

Shock propagation is studied – among many other fields – in aerodynamics and rocket propulsion, where sufficiently fast flows occur.

The Equations in One Spatial Dimension

For certain problems, especially when used to analyse compressible flow in a duct or in case the flow is cylindrically or spherically symmetric, the one-dimensional Euler equations are a useful first approximation. Generally, the Euler equations are solved by Riemann's method of characteristics. This involves finding curves in plane of independent variables (i.e., x and t) along which partial differential equations (PDE's) degenerate into ordinary differential equations (ODE's). Numerical solutions of the Euler equations rely heavily on the method of characteristics.

Hydraulics

Hydraulics is a topic in applied science and engineering dealing with the mechanical properties of liquids. Fluid mechanics provides the theoretical foundation for hydraulics, which focuses on the engineering uses of fluid properties. In fluid power, hydraulics is used for the generation, control, and transmission of power by the use of pressurised liquids. Hydraulic topics range through most science and engineering disciplines, and cover concepts such as pipe flow, dam design, fluidics and fluid control circuitry, pumps, turbines, hydropower, computational fluid dynamics, flow measurement, river channel beahviour and erosion.

Free surface hydraulics is the branch of hydraulics dealing with free surface flow, such as occurring in rivers, canals, lakes, estuaries and seas. Its sub-field open channel flow studies the flow in open channels.

The word "hydraulics" originates from the Greek word *hydraulikos* which in turn originates from *hydor*, Greek for water and *aulos*, meaning pipe.

Ancient and Medieval Era

Early uses of water power date back to Mesopotamia and ancient Egypt, where irrigation has been used since the 6th millennium BC and water clocks had been used since the early 2nd millennium BC. Other early examples of water power include the Qanat system in ancient Persia and the Turpan water system in ancient China.

Greek / Hellenistic World

Greeks continued and sophisticated the construction of water and hydraulic power systems. An example is the construction by Eupalinos, under a public contract, of a watering channel for Samos. An early example of the usage of hydraulic wheel, probably the earliest in Europe, is the Perachora wheel (3rd c. BC).

Notable is the construction of the first hydraulic automata by Ctesibius (flourished c. 270 BC) and Hero of Alexandria (c. 10–80 AD). Hero describes a number of working machines using hydraulic power, such as the force pump, which is known from many Roman sites as having been used for raising water and in fire engines.

China

In ancient China there was Sunshu Ao (6th century BC), Ximen Bao (5th century BC), Du Shi (circa 31 AD), Zhang Heng (78-139 AD), and Ma Jun (200-265 AD), while medieval China had Su Song (1020-1101 AD) and Shen Kuo (1031–1095). Du Shi employed a waterwheel to power the bellows of a blast furnace producing cast iron. Zhang Heng was the first to employ hydraulics to provide motive power in rotating an armillary sphere for astronomical observation.

Sri Lanka

In ancient Sri Lanka, hydraulics were widely used in the ancient kingdoms of Anuradhapura and Polonnaruwa. The discovery of the principle of the valve tower, or valve pit, for regulating the escape of water is credited to ingenuity more than 2,000 years ago. By the first century A.D, several large-scale irrigation works had been completed. Macro- and micro-hydraulics to provide for domestic horticultural and agricultural needs, surface drainage and erosion control, ornamental and recreational water courses and retaining structures and also cooling systems were in place in Sigiriya, Sri Lanka. The coral on the massive rock at the site includes cisterns for collecting water.

Innovations in Ancient Rome

In Ancient Rome many different hydraulic applications were developed, including public water supplies, innumerable aqueducts, power using watermills and hydraulic mining. They were among the first to make use of the siphon to carry water across valleys, and used hushing on a large scale to prospect for and then extract metal ores. They used lead widely in plumbing systems for domestic and public supply, such as feeding thermae.

Hydraulic mining was used in the gold-fields of northern Spain, which was conquered by Augustus in 25 BC. The alluvial gold-mine of Las Medulas was one of the largest of their mines. It was worked by at least 7 long aqueducts, and the water streams were used to erode the soft deposits, and then wash the tailings for the valuable gold content.

Modern era (c. 1600–1870)

Benedetto Castelli: In 1619 Benedetto Castelli (1576-1578–1643), a student of Galileo Galilei, published the book *Della Misura dell'Acque Correnti* or "On the Measurement of Running Waters", one of the foundations of modern hydrodynamics. He served as a chief consultant to the Pope on hydraulic projects, i.e., management of rivers in the Papal States, beginning in 1626.

Blaise Pascal

Blaise Pascal (1623–1662) studied fluid hydrodynamics and hydrostatics, centred on the principles of hydraulic fluids. His inventions include the hydraulic press, which multiplied a smaller force acting on a larger area into the application of a larger force totalled over a smaller area, transmitted through the same pressure (or same change of pressure) at both locations.

Pascal's law or principle states that for an incompressible fluid at rest, the difference in pressure is proportional to the difference in height and this difference remains the same whether or not the overall pressure of the fluid is changed by applying an external force. This implies that by increasing the pressure at any point in a confined fluid, there is an equal increase at every other point in the container, i.e., any change in pressure applied at any point of the fluid is transmitted undiminished throughout the fluids.

Jean Louis Marie Poiseuille

A French physician, Poiseuille researched the flow of blood through the body and discovered an important law governing the rate of flow with the diametre of the tube in which flow occurred.

In the UK

Several cities developed city-wide hydraulic power networks in the 19th century, to operate machinery such as lifts, cranes, capstans and the like. Joseph Bramah was an early innovator and William Armstrong perfected the apparatus for power delivery on an industrial scale.

In London, the London Hydraulic Power Company was a major supplier its pipes serving large parts of the West End of London, City and the Docks, but there were schemes restricted to single enterprises such as docks and railway goods yards.

Venturi Effect

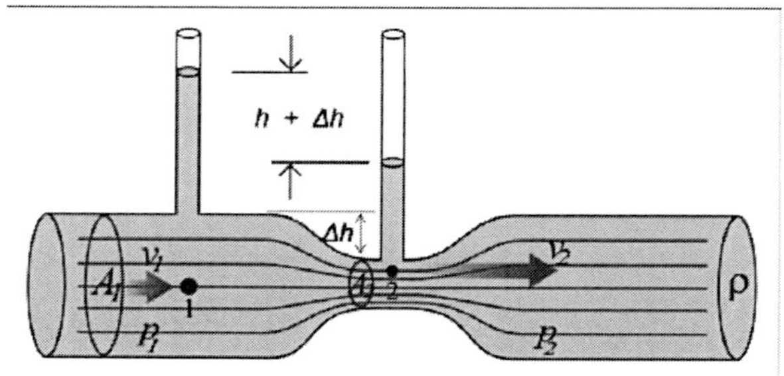

Figure: The pressure at "1" is higher than at "2", and the fluid speed at "1" is lower than at "2", because the cross-sectional area at "1" is greater than at "2".

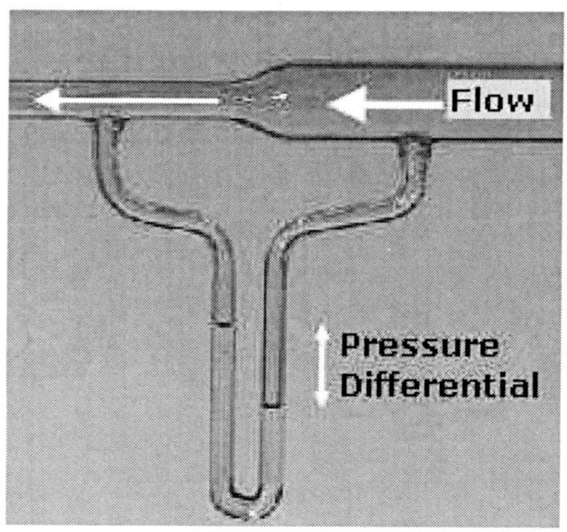

Figure: A flow of air through a venturi metre, showing the columns connected in a U-shape (a manometre) and partially filled with water. The metre is "read" as a differential pressure head in cm or inches of water.

Figure: Flow in a Venturi tube

The Venturi effect is the reduction in fluid pressure that results when a fluid flows through a constricted section of pipe. The Venturi effect is named after Giovanni Battista Venturi (1746–1822), an Italian physicist.

Background

The Venturi effect is a jet effect; as with a funnel the velocity of the fluid increases as the cross sectional area decreases, with the static pressure correspondingly decreasing. According to the laws governing fluid dynamics, a fluid's velocity must increase as it passes through a constriction to satisfy the principle of continuity, while its pressure must decrease to satisfy the principle of conservation of mechanical energy.

Thus any gain in kinetic energy a fluid may accrue due to its increased velocity through a constriction is negated by a drop in pressure. An equation for the drop in pressure due to the Venturi effect may be derived from a combination of Bernoulli's principle and the continuity equation.

The limiting case of the Venturi effect is when a fluid reaches the state of choked flow, where the fluid velocity approaches the local speed of sound. In choked flow the mass flow rate will not increase with a further decrease in the downstream pressure environment.

However, mass flow rate for a compressible fluid can increase with increased upstream pressure, which will increase the density of the fluid through the constriction (though the velocity will remain constant).

This is the principle of operation of a de Laval nozzle. Increasing source temperature will also increase the local sonic velocity, thus allowing for increased mass flow rate.

Referring to the diagram to the right, using Bernoulli's equation in the special case of incompressible flows (such as the flow of water or other liquid, or low speed flow of gas), the theoretical pressure drop at the constriction is given by:

$$p_1 - p_2 = \frac{\rho}{2}(v_2^2 - v_1^2)$$

where is the density of the fluid, is the (slower) fluid velocity where the pipe is wider, is the (faster) fluid velocity where the pipe is narrower. This assumes the flowing fluid (or other substance) is not significantly compressible-even though pressure varies, the density is assumed to remain approximately constant.

Experimental Apparatus

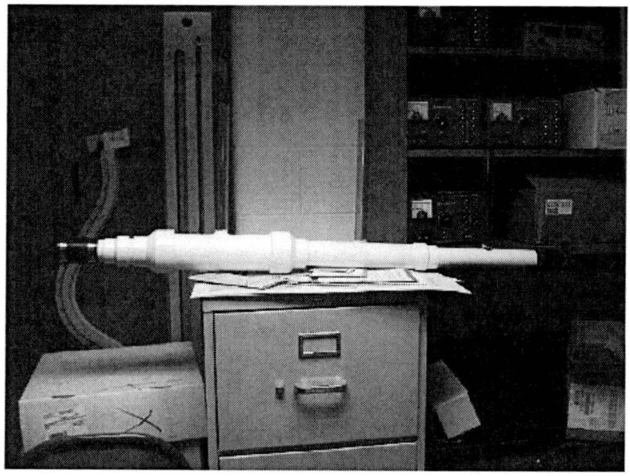

Figure: Venturi tube demonstration apparatus built out of PVC pipe and operated with a vacuum pump

Venturi Tubes

The simplest apparatus, as shown in the photograph and diagram, is a tubular setup known as a Venturi tube or simply a venturi. Fluid flows through a length of pipe of varying diametre. To avoid undue drag, a Venturi tube typically has an entry cone of 30 degrees and an exit cone of 5 degrees. To account for the assumption of an inviscid fluid a coefficient of discharge is often introduced, which generally has a value of 0.98.

Orifice Plate

Venturi tubes are more expensive to construct than a simple orifice plate which uses the same principle as a tubular scheme, but the orifice plate causes significantly more permanent energy loss.

Instrumentation and Measurement

Venturis are used in industrial and in scientific laboratories for measuring the flow of liquids.

Flow Rate

A venturi can be used to measure the volumetric flow rate Q. Since

$$\begin{cases} Q = v_1 A_1 = v_2 A_2 \\ p_1 - p_2 = \dfrac{\rho}{2}(v_2^2 - v_1^2), \end{cases}$$

then

$$Q = A_1 \sqrt{\frac{2(p_1 - p_2)}{\rho\left(\left(\frac{A_1}{A_2}\right)^2 - 1\right)}} = A_2 \sqrt{\frac{2(p_1 - p_2)}{\rho\left(1 - \left(\frac{A_2}{A_1}\right)^2\right)}}.$$

A venturi can also be used to mix a liquid with a gas. If a pump forces the liquid through a tube connected to a system consisting of a venturi to increase the liquid speed (the diametre decreases), a short piece of tube with a small hole in it, and last a venturi that decreases speed (so the pipe gets wider again), the gas will be sucked in through the small hole because of changes in pressure. At the end of the system, a mixture of liquid and gas will appear.

Differential Pressure

As fluid flows through a venturi, the expansion and compression of the fluids cause the pressure inside the venturi to change. This principle can be used in metrology for gauges calibrated for differential pressures. This type of pressure measurement may be more convenient, for example, to measure fuel or combustion pressures in jet or rocket engines.

Examples

The Venturi effect may be observed or used in the following:

- Cargo eductors on oil product and chemical ship tankers
- Inspirators that mix air and flammable gas in grills, gas stoves, Bunsen burners and airbrushes
- Water aspirators that produce a partial vacuum using the kinetic energy from the faucet water pressure
- Steam siphons using the kinetic energy from the steam pressure to create a partial vacuum
- Atomizers that disperse perfume or spray paint (i.e. from a spray gun).
- Foam firefighting nozzles and extinguishers
- Carburettors that use the effect to suck gasoline into an engine's intake air stream
- Wine aerators, used to infuse air into wine as it is poured into a glass
- The capillaries of the human circulatory system, where it indicates aortic regurgitation

- Aortic insufficiency is a chronic heart condition that occurs when the aortic valve's initial large stroke volume is released and the Venturi effect draws the walls together, which obstructs blood flow, which leads to a Pulsus Bisferiens.
- Protein skimmers (filtration devices for saltwater aquaria)
- In automated pool cleaners that use pressure-side water flow to collect sediment and debris
- The barrel of the modern-day clarinet, which uses a reverse taper to speed the air down the tube, enabling better tone, response and intonation
- Compressed air operated industrial vacuum cleaners
- Venturi scrubbers used to clean flue gas emissions
- Injectors (also called ejectors) used to add chlorine gas to water treatment chlorination systems
- Steam injectors use the Venturi effect and the latent heat of evapouration to deliver feed water to a steam locomotive boiler.
- Sand blasters used to draw fine sand in and mix it with air
- Emptying bilge water from a moving boat through a small waste gate in the hull—the air pressure inside the moving boat is greater than the water sliding by beneath
- A scuba diving regulator to assist the flow of air once it starts flowing
- Modern vapourisers to optimise efficiency
- In Venturi masks used in medical oxygen therapy
- In recoilless rifles to decrease the recoil of firing
- Ventilators
- The diffuser on an automobile
- Large cities where wind is forced between buildings
- The leadpipe of a trombone, affecting the timbre
- Foam proportioners used to induct fire fighting foam concentrate into fire protection systems

A simple way to demonstrate the Venturi effect is to squeeze and release a flexible hose in which fluid is flowing: the partial vacuum produced in the constriction is sufficient to keep the hose collapsed.

Venturi tubes are also used to measure the speed of a fluid, by measuring pressure changes at different segments of the device. Placing a liquid in a U-shaped tube and connecting the ends of the tubes to both ends of a Venturi is all that is needed. When the fluid flows though the Venturi the pressure in the two ends of the tube will differ, forcing the liquid to the "low pressure" side. The amount of that move can be calibrated to the speed of the fluid flow.

Inviscid Flow

An inviscid flow is the flow of an ideal fluid that is assumed to have no viscosity. In fluid dynamics there are problems that are easily solved by using the simplifying assumption of an inviscid flow.

The flow of fluids with low values of viscosity agree closely with inviscid flow everywhere except close to the fluid boundary where the boundary layer plays a significant role.

Reynolds Number

The assumption of inviscid flow is generally valid where viscous forces are small in comparison to inertial forces. Such flow situations can be identified as flows with a Reynolds number much greater than one. The assumption that viscous forces are negligible can be used to simplify the Navier-Stokes solution to the Euler equations.

The Euler equation governing inviscid flow is:

$$\rho\left(\frac{\partial}{\partial t}+\mathbf{u}\cdot\nabla\right)\mathbf{u}+\nabla p=0$$

which is admittedly the Newton's second law applied on a flowing infinitesimal volume element. In the steady-state case, combined with the continuity equation of mass, can be solved using potential flow theory.

Problems with the Inviscid-flow Model

While throughout much of a flow-field the effect of viscosity may be very small, a number of factors make the assumption of negligible viscosity invalid in many cases. Viscosity cannot be neglected near fluid boundaries because of the presence of a boundary layer, which enhances the effect of even a small amount of viscosity. Turbulence is also observed in some high-Reynolds-number flows, and is a process through which energy is transferred to increasingly small scales of motion until it is dissipated by viscosity.

Navier–Stokes Equations

In physics, the Navier–Stokes equations, named after Claude-Louis Navier and George Gabriel Stokes, describe the motion of fluid substances. These equations arise from applying Newton's second law to fluid motion, together with the assumption that the fluid stress is the sum of a diffusing viscous term (proportional to the gradient of velocity), plus a pressure term.

The equations are useful because they describe the physics of many things of academic and economic interest. They may be used to model the weather, ocean currents, water flow in a pipe and air flow around a wing. The Navier–Stokes equations in their full and simplified forms help with the design of aircraft and cars, the study of blood flow, the design of power stations, the analysis of pollution, and many other things. Coupled with Maxwell's equations they can be used to model and study magnetohydrodynamics.

The Navier–Stokes equations are also of great interest in a purely mathematical sense. Somewhat surprisingly, given their wide range of practical uses, mathematicians have not yet proven that in three dimensions solutions always exist (existence), or that if they do exist, then they do not contain any singularity (smoothness). These are called the Navier–Stokes existence and smoothness problems. The Clay Mathematics Institute has called this one of the seven most important open problems in mathematics and has offered a US$1,000,000 prize for a solution or a counter-example.

Velocity Field

The Navier–Stokes equations dictate not position but rather velocity. A solution of the Navier–Stokes equations is called a velocity field or flow field, which is a description of the velocity of the fluid at a given point in space and time. Once the velocity field is solved for, other quantities of interest (such as flow rate or drag force) may be found. This is different from what one normally sees in classical mechanics, where solutions are typically trajectories of position of a particle or deflection of a continuum. Studying velocity instead of position makes more sense for a fluid; however for visualisation purposes one can compute various trajectories.

Properties

Nonlinearity: The Navier–Stokes equations are nonlinear partial differential equations in almost every real situation. In some cases, such as one-dimensional flow and Stokes flow (or creeping flow), the equations can be simplified to linear equations. The nonlinearity makes most problems difficult or impossible to solve and is the main contributor to the turbulence that the equations model.

The nonlinearity is due to convective acceleration, which is an acceleration associated with the change in velocity over position. Hence, any convective flow, whether turbulent or not, will involve nonlinearity. An example of convective but laminar (nonturbulent) flow would be the passage of a viscous fluid (for example, oil) through a small converging nozzle. Such flows, whether exactly solvable or not, can often be thoroughly studied and understood.

Turbulence

Turbulence is the time dependent chaotic beahviour seen in many fluid flows. It is generally believed that it is due to the inertia of the fluid as a whole: the culmination of time dependent and convective acceleration; hence flows where inertial effects are small tend to be laminar (the Reynolds number quantifies how much the flow is affected by inertia). It is believed, though not known with certainty, that the Navier–Stokes equations describe turbulence properly.

The numerical solution of the Navier–Stokes equations for turbulent flow is extremely difficult, and due to the significantly different mixing-length scales that are involved in turbulent flow, the stable solution of this requires such a fine mesh resolution that the computational time becomes significantly infeasible for calculation. Attempts to solve turbulent flow using a laminar solver typically result in a time-unsteady solution, which fails to converge appropriately. To counter this, time-averaged equations such as the Reynolds-averaged Navier–Stokes equations (RANS), supplemented with turbulence models, are used in practical computational fluid dynamics (CFD) applications when modelling turbulent flows. Some models include the Spalart-Allmaras, k-ω (k-omega), k-ε (k-epsilon), and SST models which add a variety of additional equations to bring closure to the RANS equations. Another technique for solving numerically the Navier–Stokes equation is the Large eddy simulation (LES). This approach is computationally more expensive than the RANS method (in time and computer memory), but produces better results since the larger turbulent scales are explicitly resolved.

Applicability

Together with supplemental equations (for example, conservation of mass) and well formulated boundary conditions, the Navier–Stokes equations seem to model fluid motion accurately; even turbulent flows seem (on average) to agree with real world observations.

The Navier–Stokes equations assume that the fluid being studied is a continuum (it is infinitely divisible and not composed of particles such as atoms or molecules), and is not moving at relativistic velocities. At very small scales or under extreme conditions, real fluids made out of discrete molecules will produce results different from the continuous fluids modelled by the Navier–Stokes equations. Depending on the Knudsen number of the problem, statistical mechanics or possibly even molecular dynamics may be a more appropriate approach.

Another limitation is simply the complicated nature of the equations. Time tested formulations exist for common fluid families, but the application of the Navier–Stokes equations to less common families tends to result in very complicated formulations which are an area of current research. For this reason, these equations are usually written for Newtonian fluids. Studying such fluids is "simple" because the viscosity model ends up being linear; truly general models for the flow of other kinds of fluids (such as blood) do not, as of 2012, exist.

Derivation and Description

The derivation of the Navier–Stokes equations begins with an application of Newton's second law: conservation of momentum (often alongside mass and energy conservation) being written for an arbitrary portion of the fluid. In an inertial frame of reference, the general form of the equations of fluid motion is:

Navier–Stokes equations *(general)*

$$\rho\left(\frac{\partial \mathbf{v}}{\partial t}+\mathbf{v}\cdot\nabla\mathbf{v}\right)=-\nabla p+\nabla\cdot\mathbb{T}+\mathbf{f},$$

where v is the flow velocity, ρ is the fluid density, p is the pressure, is the (deviatoric) stress tensor, and f represents body forces (per unit volume) acting on the fluid and " is the del operator. This is a statement of the conservation of momentum in a fluid and it is an application of Newton's second law to a continuum; in fact this equation is applicable to any non-relativistic continuum and is known as the Cauchy momentum equation.

This equation is often written using the material derivative *D*v/*Dt*, making it more apparent that this is a statement of Newton's second law:

$$\rho\frac{D\mathbf{v}}{Dt}=-\nabla p+\nabla\cdot\mathbb{T}+\mathbf{f}.$$

The left side of the equation describes acceleration, and may be composed of time dependent or convective effects (also the effects of non-inertial coordinates if present). The right side of the equation is in effect a summation of body forces (such as gravity) and divergence of stress (pressure and shear stress).

Convective Acceleration

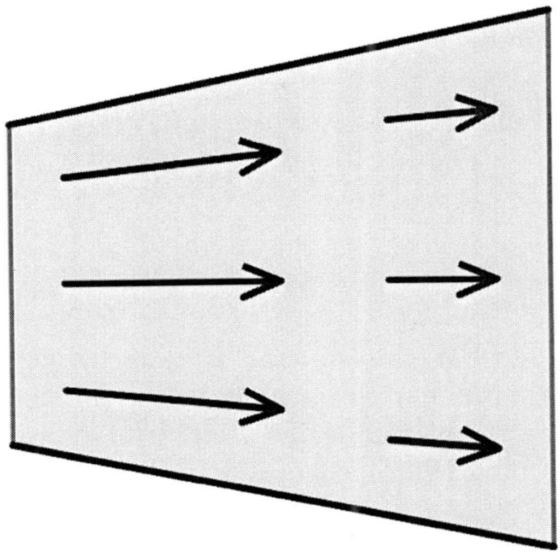

Figure: An example of convection. Though the flow may be steady (time independent), the fluid decelerates as it moves down the diverging duct (assuming incompressible or subsonic compressible flow), hence there is an acceleration happening over position.

A very significant feature of the Navier–Stokes equations is the presence of convective acceleration: the effect of time independent acceleration of a fluid with respect to space. While individual fluid particles are indeed experiencing time dependent acceleration, the convective acceleration of the flow field is a spatial effect, one example being fluid speeding up in a nozzle. Convective acceleration is represented by the nonlinear quantity:

$$\mathbf{v}\cdot\nabla\mathbf{v},$$

which may be interpreted either as $(\mathbf{v}\cdot\nabla)\mathbf{v}$ or as $(\mathbf{v}\cdot\nabla)\mathbf{v}$ with $\nabla\mathbf{v}$ the tensor derivative of the velocity vector v Both interpretations give the same result, independent of the coordinate system — provided V is interpreted as the covariant derivative.

Interpretation as $(v\cdot\nabla)v$

The convection term is often written as

$$(\mathbf{v}\cdot\nabla)\mathbf{v},$$

where the advection operator $\mathbf{v}\cdot\nabla$ is used. Usually this representation is preferred as it is simpler than the one in terms of the tensor derivative $\nabla\mathbf{v}$.

Interpretation as v ·(∇v)

Here $\nabla\mathbf{v}$. is the tensor derivative of the velocity vector, equal in Cartesian coordinates to the component by component gradient. The convection term may, by a vector calculus identity, be expressed without a tensor derivative:

$$\mathbf{v}\cdot\nabla\mathbf{v} = \nabla\left(\frac{|\mathbf{v}|^2}{2}\right) + (\nabla\times\mathbf{v})\times\mathbf{v}.$$

The form has use in irrotational flow, where the curl of the velocity (called vorticity) $\omega = \nabla\times\mathbf{v}$ is equal to zero.

Regardless of what kind of fluid is being dealt with, convective acceleration is a nonlinear effect. Convective acceleration is present in most flows (exceptions include one-dimensional incompressible flow), but its dynamic effect is disregarded in creeping flow (also called Stokes flow).

Stresses

The effect of stress in the fluid is represented by the ∇p and $\nabla\cdot\mathbb{T}$ terms; these are gradients of surface forces, analogous to stresses in a solid ∇p is called the pressure gradient and arises from the isotropic part of the stress tensor. This part is given by normal stresses that turn up in almost all situations, dynamic or not. The anisotropic part of the stress tensor gives rise to $\nabla\cdot\mathbb{T}$, which conventionally describes viscous forces; for incompressible flow, this T is only a shear effect. Thus, is the deviatoric stress tensor, and the stress tensor is equal to:

$$\sigma = -p\mathbb{I} + \mathbb{T}$$

where is the 3×3 identity matrix. Interestingly, only the *gradient* of pressure matters, not the pressure itself. The effect of the pressure gradient is that fluid flows from high pressure to low pressure.

The stress terms p and T are yet unknown, so the general form of the equations of motion is not usable to solve problems. Besides the equations of motion—Newton's second law—a force model is needed relating the stresses to the fluid motion. For this reason, assumptions on the specific beahviour of a fluid are made (based on natural observations) and applied in order to specify the stresses in terms of the other flow variables, such as velocity and density.

The Navier–Stokes equations result from the following assumptions on the deviatoric stress tensor T:

- the deviatoric stress vanishes for a fluid at rest, and – by Galilean invariance – also does not depend directly on the flow velocity itself, but only on spatial derivatives of the flow velocity
- in the Navier–Stokes equations, the deviatoric stress is expressed as the product of the tensor gradient $\nabla \mathbf{v}$ of the flow velocity with a viscosity tensor A, i.e. : $\mathbb{T} = \mathbb{A}(\nabla \mathbf{v})$
- the fluid is assumed to be isotropic, as valid for gases and simple liquids, and consequently A is an isotropic tensor; furthermore, since the deviatoric stress tensor is symmetric, it turns out that it can be expressed in terms of two scalar dynamic viscosities μ and μ-: $\mathbb{T} = 2\mu\mathbb{E} + \mu''\Delta\mathbb{I}$, where $\mathbb{E} = \frac{1}{2}(\nabla \mathbf{v}) + \frac{1}{2}(\nabla \mathbf{v})^{T}$ is the rate-of-strain tensor and $\Delta = \nabla \cdot \mathbf{v}$ is the rate of expansion of the flow
- the deviatoric stress tensor has zero trace, so for a three-dimensional flow $2\mu + 3\mu$- $= 0$

As a result, in the Navier–Stokes equations the deviatoric stress tensor has the following form:

$$\mathbb{T} = 2\mu\left(\mathbb{E} - \tfrac{1}{3}\Delta\mathbb{I}\right),$$

with the quantity between brackets the non-isotropic part of the rate-of-strain tensor E. The dynamic viscosity μ does not need to be constant – in general it depends on conditions like temperature and pressure, and in turbulence modelling the concept of eddy viscosity is used to approximate the average deviatoric stress.

The pressure p is modelled by use of an equation of state. For the special case of an incompressible flow, the pressure constrains the flow in such a way that the volume of fluid elements is constant: isochoric flow resulting in a solenoidal velocity field with $\nabla \cdot \mathbf{v} = 0$.

Other Forces

The vector field represents body forces. Typically these consist of only gravity forces, but may include other types (such as electromagnetic forces). In a non-inertial coordinate system, other "forces" such as that associated with rotating coordinates may be inserted.

Often, these forces are so-called conservative forces and may be represented as the gradient of some scalar quantity F, with $\mathbf{f} = \nabla F$.

Gravity in the z direction, for example, is the gradient of $-\rho gz$. Since pressure shows up only as a gradient, this implies that solving a problem without any such body force can be mended to include the body force by using a modified pressure $p_m = p - F$. The pressure and force terms on the right hand side of the Navier–Stokes equation become

$$-\nabla p + \mathbf{f} = -\nabla p + \nabla F = -\nabla\left(p - F\right) = -\nabla p_m.$$

Other Equations

The Navier–Stokes equations are strictly a statement of the conservation of momentum. In order to fully describe fluid flow, more information is needed (how much depends on the assumptions made). This additional information may include boundary data (no-slip, capillary surface, etc.), the conservation of mass, the conservation of energy, and/or an equation of state.

Regardless of the flow assumptions, a statement of the conservation of mass is generally necessary. This is achieved through the mass continuity equation, given in its most general form as:

$$\frac{\partial \rho}{\partial t} + \nabla \cdot (\rho \mathbf{v}) = 0$$

or, using the substantive derivative:

$$\frac{D\rho}{Dt} + \rho(\nabla \cdot \mathbf{v}) = 0.$$

Incompressible Flow of Newtonian Fluids

A simplification of the resulting flow equations is obtained when considering an incompressible flow of a Newtonian fluid. The assumption of incompressibility rules out the possibility of sound or shock waves to occur; so this simplification is invalid if these phenomena are important. The incompressible flow assumption typically holds well even when dealing with a "compressible" fluid — such as air at room temperature — at low Mach numbers (even when flowing up to about Mach 0.3). Taking the incompressible flow assumption into account and assuming constant viscosity, the Navier–Stokes equations will read, in vector form:

Navier–Stokes equations *(Incompressible flow)*

$$\rho\left(\frac{\partial \mathbf{v}}{\partial t} + \mathbf{v} \cdot \nabla \mathbf{v}\right) = -\nabla p + \mu \nabla^2 \mathbf{v} + \mathbf{f}.$$

Here f represents “other” body forces (forces per unit volume), such as gravity or centrifugal force. The shear stress term becomes the useful quantity (is the vector Laplacian) when the fluid is assumed incompressible, homogeneous and Newtonian, where is the (constant) dynamic viscosity.

It’s well worth observing the meaning of each term (compare to the Cauchy momentum equation):

$$\overbrace{\rho\Big(\underbrace{\frac{\partial \mathbf{v}}{\partial t}}_{\substack{\text{Unsteady}\\ \text{acceleration}}} + \underbrace{\mathbf{v}\cdot\nabla\mathbf{v}}_{\substack{\text{Convective}\\ \text{acceleration}}}\Big)}^{\text{Inertia (per volume)}} = \overbrace{\underbrace{-\nabla p}_{\substack{\text{Pressure}\\ \text{gradient}}} + \underbrace{\mu\nabla^2\mathbf{v}}_{\text{Viscosity}}}^{\text{Divergence of stress}} + \underbrace{\mathbf{f}}_{\substack{\text{Other}\\ \text{body}\\ \text{forces}}} .$$

Note that only the convective terms are nonlinear for incompressible Newtonian flow. The convective acceleration is an acceleration caused by a (possibly steady) change in velocity over *position*, for example the speeding up of fluid entering a converging nozzle.

Though individual fluid particles are being accelerated and thus are under unsteady motion, the flow field (a velocity distribution) will not necessarily be time dependent.

Another important observation is that the viscosity is represented by the vector Laplacian of the velocity field (interpreted here as the difference between the velocity at a point and the mean velocity in a small volume around). This implies that Newtonian viscosity is diffusion of momentum, this works in much the same way as the diffusion of heat seen in the heat equation (which also involves the Laplacian).

If temperature effects are also neglected, the only “other” equation (apart from initial/boundary conditions) needed is the mass continuity equation. Under the incompressible assumption, density is a constant and it follows that the equation will simplify to:

$$\nabla\cdot\mathbf{v} = 0.$$

This is more specifically a statement of the conservation of volume.

These equations are commonly used in 3 coordinates systems: Cartesian, cylindrical, and spherical. While the Cartesian equations seem to follow directly from the vector equation above, the vector form of the Navier–Stokes equation involves some tensor calculus which means that writing it in other coordinate systems is not as simple as doing so for scalar equations (such as the heat equation).

Cartesian Coordinates

Writing the vector equation explicitly,

$$\rho\left(\frac{\partial u}{\partial t}+u\frac{\partial u}{\partial x}+v\frac{\partial u}{\partial y}+w\frac{\partial u}{\partial z}\right)=-\frac{\partial p}{\partial x}+\mu\left(\frac{\partial^2 u}{\partial x^2}+\frac{\partial^2 u}{\partial y^2}+\frac{\partial^2 u}{\partial z^2}\right)+\rho g_x$$

$$\rho\left(\frac{\partial v}{\partial t}+u\frac{\partial v}{\partial x}+v\frac{\partial v}{\partial y}+w\frac{\partial v}{\partial z}\right)=-\frac{\partial p}{\partial y}+\mu\left(\frac{\partial^2 v}{\partial x^2}+\frac{\partial^2 v}{\partial y^2}+\frac{\partial^2 v}{\partial z^2}\right)+\rho g_y$$

$$\rho\left(\frac{\partial w}{\partial t}+u\frac{\partial w}{\partial x}+v\frac{\partial w}{\partial y}+w\frac{\partial w}{\partial z}\right)=-\frac{\partial p}{\partial z}+\mu\left(\frac{\partial^2 w}{\partial x^2}+\frac{\partial^2 w}{\partial y^2}+\frac{\partial^2 w}{\partial z^2}\right)+\rho g_z$$

The continuity equation reads:

$$\frac{\partial \rho}{\partial t}+\frac{\partial(\rho u)}{\partial x}+\frac{\partial(\rho v)}{\partial y}+\frac{\partial(\rho w)}{\partial z}=0.$$

When the flow is at steady-state, does not change with respect to time. The continuity equation is reduced to:

$$\frac{\partial(\rho u)}{\partial x}+\frac{\partial(\rho v)}{\partial y}+\frac{\partial(\rho w)}{\partial z}=0.$$

When the flow is incompressible, is constant and does not change with respect to space. The continuity equation is reduced to:

$$\frac{\partial u}{\partial x}+\frac{\partial v}{\partial y}+\frac{\partial w}{\partial z}=0.$$

The velocity components (the dependent variables to be solved for) are typically named u, v, w. This system of four equations comprises the most commonly used and studied form. Though comparatively more compact than other representations, this is still a nonlinear system of partial differential equations for which solutions are difficult to obtain.

Cylindrical Coordinates

A change of variables on the Cartesian equations will yield the following momentum equations for r, ϕ, and z:

$$r:\ \rho\left(\frac{\partial u_r}{\partial t}+u_r\frac{\partial u_r}{\partial r}+\frac{u_\phi}{r}\frac{\partial u_r}{\partial \phi}+u_z\frac{\partial u_r}{\partial z}-\frac{u_\phi^2}{r}\right)$$

$$= -\frac{\partial p}{\partial r} + \mu\left[\frac{1}{r}\frac{\partial}{\partial r}\left(r\frac{\partial u_r}{\partial r}\right) + \frac{1}{r^2}\frac{\partial^2 u_r}{\partial \phi^2} + \frac{\partial^2 u_r}{\partial z^2} - \frac{u_r}{r^2} - \frac{2}{r^2}\frac{\partial u_\phi}{\partial \phi}\right] + \rho g_r$$

$$\phi:\ \rho\left(\frac{\partial u_\phi}{\partial t} + u_r\frac{\partial u_\phi}{\partial r} + \frac{u_\phi}{r}\frac{\partial u_\phi}{\partial \phi} + u_z\frac{\partial u_\phi}{\partial z} + \frac{u_r u_\phi}{r}\right)$$

$$= -\frac{1}{r}\frac{\partial p}{\partial \phi} + \mu\left[\frac{1}{r}\frac{\partial}{\partial r}\left(r\frac{\partial u_\phi}{\partial r}\right) + \frac{1}{r^2}\frac{\partial^2 u_\phi}{\partial \phi^2} + \frac{\partial^2 u_\phi}{\partial z^2} + \frac{2}{r^2}\frac{\partial u_r}{\partial \phi} - \frac{u_\phi}{r^2}\right] + \rho g_\phi$$

$$z:\ \rho\left(\frac{\partial u_z}{\partial t} + u_r\frac{\partial u_z}{\partial r} + \frac{u_\phi}{r}\frac{\partial u_z}{\partial \phi} + u_z\frac{\partial u_z}{\partial z}\right)$$

$$= -\frac{\partial p}{\partial z} + \mu\left[\frac{1}{r}\frac{\partial}{\partial r}\left(r\frac{\partial u_z}{\partial r}\right) + \frac{1}{r^2}\frac{\partial^2 u_z}{\partial \phi^2} + \frac{\partial^2 u_z}{\partial z^2}\right] + \rho g_z.$$

The gravity components will generally not be constants, however for most applications either the coordinates are chosen so that the gravity components are constant or else it is assumed that gravity is counteracted by a pressure field (for example, flow in horizontal pipe is treated normally without gravity and without a vertical pressure gradient). The continuity equation is:

$$\frac{\partial \rho}{\partial t} + \frac{1}{r}\frac{\partial}{\partial r}(\rho r u_r) + \frac{1}{r}\frac{\partial(\rho u_\phi)}{\partial \phi} + \frac{\partial(\rho u_z)}{\partial z} = 0.$$

This cylindrical representation of the incompressible Navier–Stokes equations is the second most commonly seen (the first being Cartesian above).

Cylindrical coordinates are chosen to take advantage of symmetry, so that a velocity component can disappear. A very common case is axisymmetric flow with the assumption of no tangential velocity ($u_\phi = 0$), and the remaining quantities are independent of ϕ:

$$\rho\left(\frac{\partial u_r}{\partial t} + u_r\frac{\partial u_r}{\partial r} + u_z\frac{\partial u_r}{\partial z}\right) = -\frac{\partial p}{\partial r} + \mu\left[\frac{1}{r}\frac{\partial}{\partial r}\left(r\frac{\partial u_r}{\partial r}\right) + \frac{\partial^2 u_r}{\partial z^2} - \frac{u_r}{r^2}\right] + \rho g_r$$

$$\rho\left(\frac{\partial u_z}{\partial t} + u_r\frac{\partial u_z}{\partial r} + u_z\frac{\partial u_z}{\partial z}\right) = -\frac{\partial p}{\partial z} + \mu\left[\frac{1}{r}\frac{\partial}{\partial r}\left(r\frac{\partial u_z}{\partial r}\right) + \frac{\partial^2 u_z}{\partial z^2}\right] + \rho g_z$$

$$\frac{1}{r}\frac{\partial}{\partial r}(r u_r) + \frac{\partial u_z}{\partial z} = 0.$$

Spherical Coordinates

In spherical coordinates, the r, ϕ, and θ momentum equations are (note the convention used: θ is polar angle, or colatitude, $0 \leq \theta \leq \pi$):

$$r:\ \rho\left(\frac{\partial u_r}{\partial t}+u_r\frac{\partial u_r}{\partial r}+\frac{u_\phi}{r\sin(\theta)}\frac{\partial u_r}{\partial \phi}+\frac{u_\theta}{r}\frac{\partial u_r}{\partial \theta}-\frac{u_\phi^2+u_\theta^2}{r}\right)=-\frac{\partial p}{\partial r}+\rho g_r+$$

$$\mu\left[\frac{1}{r^2}\frac{\partial}{\partial r}\left(r^2\frac{\partial u_r}{\partial r}\right)+\frac{1}{r^2\sin(\theta)^2}\frac{\partial^2 u_r}{\partial \phi^2}+\frac{1}{r^2\sin(\theta)}\frac{\partial}{\partial \theta}\left(\sin(\theta)\frac{\partial u_r}{\partial \theta}\right)-2\frac{u_r+\frac{\partial u_\theta}{\partial \theta}+u_\theta\cot(\theta)}{r^2}-\frac{2}{r^2\sin(\theta)}\frac{\partial u_\phi}{\partial \phi}\right]$$

$$\phi:\ \rho\left(\frac{\partial u_\phi}{\partial t}+u_r\frac{\partial u_\phi}{\partial r}+\frac{u_\phi}{r\sin(\theta)}\frac{\partial u_\phi}{\partial \phi}+\frac{u_\theta}{r}\frac{\partial u_\phi}{\partial \theta}+\frac{u_r u_\phi+u_\phi u_\theta\cot(\theta)}{r}\right)$$

$$=-\frac{1}{r\sin(\theta)}\frac{\partial p}{\partial \phi}+\rho g_\phi+$$

$$\mu\left[\frac{1}{r^2}\frac{\partial}{\partial r}\left(r^2\frac{\partial u_\phi}{\partial r}\right)+\frac{1}{r^2\sin(\theta)^2}\frac{\partial^2 u_\phi}{\partial \phi^2}+\frac{1}{r^2\sin(\theta)}\frac{\partial}{\partial \theta}\left(\sin(\theta)\frac{\partial u_\phi}{\partial \theta}\right)+\frac{2\sin(\theta)\frac{\partial u_r}{\partial \phi}+2\cos(\theta)\frac{\partial u_\theta}{\partial \phi}-u_\phi}{r^2\sin(\theta)^2}\right]$$

$$\theta:\ \rho\left(\frac{\partial u_\theta}{\partial t}+u_r\frac{\partial u_\theta}{\partial r}+\frac{u_\phi}{r\sin(\theta)}\frac{\partial u_\theta}{\partial \phi}+\frac{u_\theta}{r}\frac{\partial u_\theta}{\partial \theta}+\frac{u_r u_\theta-u_\phi^2\cot(\theta)}{r}\right)$$

$$=-\frac{1}{r}\frac{\partial p}{\partial \theta}+\rho g_\theta+$$

$$\mu\left[\frac{1}{r^2}\frac{\partial}{\partial r}\left(r^2\frac{\partial u_\theta}{\partial r}\right)+\frac{1}{r^2\sin(\theta)^2}\frac{\partial^2 u_\theta}{\partial \phi^2}+\frac{1}{r^2\sin(\theta)}\frac{\partial}{\partial \theta}\left(\sin(\theta)\frac{\partial u_\theta}{\partial \theta}\right)+\frac{2}{r^2}\frac{\partial u_r}{\partial \theta}-\frac{u_\theta+2\cos(\theta)\frac{\partial u_\phi}{\partial \phi}}{r^2\sin(\theta)^2}\right].$$

Mass continuity will read:

$$\frac{\partial \rho}{\partial t}+\frac{1}{r^2}\frac{\partial}{\partial r}\left(\rho r^2 u_r\right)+\frac{1}{r\sin(\theta)}\frac{\partial \rho u_\phi}{\partial \phi}+\frac{1}{r\sin(\theta)}\frac{\partial}{\partial \theta}\left(\sin(\theta)\rho u_\theta\right)=0.$$

These equations could be (slightly) compacted by, for example, factoring from $1/r^2$ the viscous terms. However, doing so would undesirably alter the structure of the Laplacian and other quantities.

Stream Function Formulation

Taking the curl of the Navier–Stokes equation results in the elimination of pressure. This is especially easy to see if 2D Cartesian flow is assumed (and no dependence of anything on z), where the equations reduce to:

$$\rho\left(\frac{\partial u}{\partial t}+u\frac{\partial u}{\partial x}+v\frac{\partial u}{\partial y}\right)=-\frac{\partial p}{\partial x}+\mu\left(\frac{\partial^2 u}{\partial x^2}+\frac{\partial^2 u}{\partial y^2}\right)+\rho g_x,$$

$$\rho\left(\frac{\partial v}{\partial t}+u\frac{\partial v}{\partial x}+v\frac{\partial v}{\partial y}\right)=-\frac{\partial p}{\partial y}+\mu\left(\frac{\partial^2 v}{\partial x^2}+\frac{\partial^2 v}{\partial y^2}\right)+\rho g_y.$$

Differentiating the first with respect to y, the second with respect to x and subtracting the resulting equations will eliminate pressure and any conservative force. Defining the stream function ψ through

$$u=\frac{\partial \psi}{\partial y} \quad ; \quad v=-\frac{\partial \psi}{\partial x}$$

results in mass continuity being unconditionally satisfied (given the stream function is continuous), and then incompressible Newtonian 2D momentum and mass conservation degrade into one equation:

$$\frac{\partial}{\partial t}\left(\nabla^2\psi\right)+\frac{\partial \psi}{\partial y}\frac{\partial}{\partial x}\left(\nabla^2\psi\right)-\frac{\partial \psi}{\partial x}\frac{\partial}{\partial y}\left(\nabla^2\psi\right)=\nu\nabla^4\psi,$$

where ∇^4 is the (2D) biharmonic operator and υ is the kinematic viscosity, $\nu=\frac{\mu}{\rho}$. We can also express this compactly using the Jacobian determinant:

$$\frac{\partial}{\partial t}\left(\nabla^2\psi\right)+\frac{\partial\left(\psi,\nabla^2\psi\right)}{\partial\left(y,x\right)}=\nu\nabla^4\psi.$$

This single equation together with appropriate boundary conditions describes 2D fluid flow, taking only kinematic viscosity as a parameter. Note that the equation for creeping flow results when the left side is assumed zero. In axisymmetric flow another stream function formulation, called the Stokes stream function, can be used to describe the velocity components of an incompressible flow with one scalar function.

Pressure-free Velocity Formulation

The incompressible Navier–Stokes equation is a differential algebraic equation, having the inconvenient feature that there is no explicit mechanism for advancing the pressure in time. Consequently, much effort has been expended to eliminate the pressure from all or part of the computational process. The stream function formulation above eliminates the pressure (in 2D) at the expense of introducing higher derivatives and elimination of the velocity, which is the primary variable of interest.

The incompressible Navier–Stokes equation is composite, the sum of two orthogonal equations,

$$\frac{\partial \mathbf{v}}{\partial t} = \Pi^S\left(-\mathbf{v}\cdot\nabla\mathbf{v} + \nu\nabla^2\mathbf{v}\right) + \mathbf{f}^S,$$

$$\rho^{-1}\nabla p = \Pi^I\left(-\mathbf{v}\cdot\nabla\mathbf{v} + \nu\nabla^2\mathbf{v}\right) + \mathbf{f}^I,$$

where Π^S and Π^I are solenoidal and irrotational projection operators satisfying $\Pi^S + \Pi^I = 1$ and $\mathbf{f}^S$ and $\mathbf{f}^I$ are the nonconservative and conservative parts of the body force. This result follows from the Helmholtz Theorem (also known as the fundamental theorem of vector calculus). The first equation is a pressureless governing equation for the velocity, while the second equation for the pressure is a functional of the velocity and is related to the pressure Poisson equation.

The explicit functional form of the projection operator in 3D is found from the Helmholtz Theorem

$$\Pi^S\,\mathbf{F}(\mathbf{r}) = \frac{1}{4\pi}\nabla\times\int\frac{\nabla'\times\mathbf{F}(\mathbf{r}')}{|\mathbf{r}-\mathbf{r}'|}dV', \quad \Pi^I = 1 - \Pi^S.$$

with a similar structure in 2D. Thus the governing equation is an integro-differential equation and not convenient for numerical computation.

An equivalent weak or variational form of the equation, proved to produce the same velocity solution as the Navier–Stokes equation, is given by,

$$(\mathbf{w}, \frac{\partial \mathbf{v}}{\partial t}) = -(\mathbf{w}, \mathbf{v}\cdot\nabla\mathbf{v}) - \nu(\nabla\mathbf{w} : \nabla\mathbf{v}) + (\mathbf{w}, \mathbf{f}^S)$$

for divergence-free test functions satisfying appropriate boundary conditions. Here, the projections are accomplished by the orthogonality of the solenoidal and irrotational function spaces. The discrete form of this is imminently suited to finite element computation of divergence-free flow. There we will be able to address the question, "How does one specify pressure-driven (Poiseuille) problems with a pressureless governing equation?"

The absence of pressure forces from the governing velocity equation demonstrates that the equation is not a dynamic one, but rather a kinematic equation where the divergence-free condition serves the role of a conservation law. This all would seem to refute the frequent statements that the incompressible pressure enforces the divergence-free condition.

Discrete Velocity

With partitioning of the problem domain and defining basis functions on the partitioned domain, the discrete form of the governing equation is,

$$\left(\mathbf{w}_i, \frac{\partial \mathbf{v}_j}{\partial t}\right) = -(\mathbf{w}_i, \mathbf{v}\cdot\nabla\mathbf{v}_j) - \nu(\nabla\mathbf{w}_i : \nabla\mathbf{v}_j) + \left(\mathbf{w}_i, \mathbf{f}^S\right).$$

It is desirable to choose basis functions which reflect the essential feature of incompressible flow – the elements must be divergence-free. While the velocity is the variable of interest, the existence of the stream function or vector potential is necessary by the Helmholtz Theorem. Further, to determine fluid flow in the absence of a pressure gradient, one can specify the difference of stream function values across a 2D channel, or the line integral of the tangential component of the vector potential around the channel in 3D, the flow being given by Stokes' Theorem. Discussion will be restricted to 2D in the following.

We further restrict discussion to continuous Hermite finite elements which have at least first-derivative degrees-of-freedom. With this, one can draw a large number of candidate triangular and rectangular elements from the plate-bending literature. These elements have derivatives as components of the gradient. In 2D, the gradient and curl of a scalar are clearly orthogonal, given by the expressions,

$$\nabla\phi = \left[\frac{\partial\phi}{\partial x}, \frac{\partial\phi}{\partial y}\right]^T, \quad \nabla\times\phi = \left[\frac{\partial\phi}{\partial y}, -\frac{\partial\phi}{\partial x}\right]^T.$$

Adopting continuous plate-bending elements, interchanging the derivative degrees-of-freedom and changing the sign of the appropriate one gives many families of stream function elements.

Taking the curl of the scalar stream function elements gives divergence-free velocity elements. The requirement that the stream function elements be continuous assures that the normal component of the velocity is continuous across element interfaces, all that is necessary for vanishing divergence on these interfaces.

Boundary conditions are simple to apply. The stream function is constant on no-flow surfaces, with no-slip velocity conditions on surfaces. Stream function differences across open channels determine the flow. No boundary conditions are necessary on open boundaries, though consistent values may be used with some problems. These are all Dirichlet conditions.

The algebraic equations to be solved are simple to set up, but of course are non-linear, requiring iteration of the linearised equations.

Similar considerations apply to three-dimensions, but extension from 2D is not immediate because of the vector nature of the potential, and there exists no simple relation between the gradient and the curl as was the case in 2D.

Pressure Recovery

Recovering pressure from the velocity field is easy. The discrete weak equation for the pressure gradient is,

$$(\mathbf{g}_i, \nabla p) = -(\mathbf{g}_i, \mathbf{v}\cdot\nabla\mathbf{v}_j) - \nu(\nabla\mathbf{g}_i : \nabla\mathbf{v}_j) + (\mathbf{g}_i, \mathbf{f}^I),$$

where the test/weight functions are irrotational. Any conforming scalar finite element may be used. However, the pressure gradient field may also be of interest. In this case one can use scalar Hermite elements for the pressure. For the test/weight functions $\mathbf{g}_i$ one would choose the irrotational vector elements obtainied from the gradient of the pressure element.

Compressible Flow of Newtonian Fluids

There are some phenomena that are closely linked with fluid compressibility. One of the obvious examples is sound. Description of such phenomena requires more general presentation of the Navier–Stokes equation that takes into account fluid compressibility. If viscosity is assumed a constant, one additional term appears, as shown here:

$$\rho\left(\frac{\partial\mathbf{v}}{\partial t} + \mathbf{v}\cdot\nabla\mathbf{v}\right) = -\nabla p + \mu\nabla^2\mathbf{v} + \left(\tfrac{1}{3}\mu + \mu^v\right)\nabla(\nabla\cdot\mathbf{v}) + \mathbf{f}$$

where μ^v is the volume viscosity coefficient, also known as *bulk viscosity*. This additional term disappears for an *incompressible fluid*, when the divergence of the flow equals zero.

Application to Specific Problems

The Navier–Stokes equations, even when written explicitly for specific fluids, are rather generic in nature and their proper application to specific problems can be very diverse. This is partly because there is an enormous variety of problems that may be modelled, ranging from as simple as the distribution of static pressure to as complicated as multiphase flow driven by surface tension.

Generally, application to specific problems begins with some flow assumptions and initial/boundary condition formulation, this may be followed by scale analysis to further simplify the problem. For example, after assuming steady, parallel, one dimensional, nonconvective pressure driven flow between parallel plates, the resulting scaled (dimensionless) boundary value problem is:

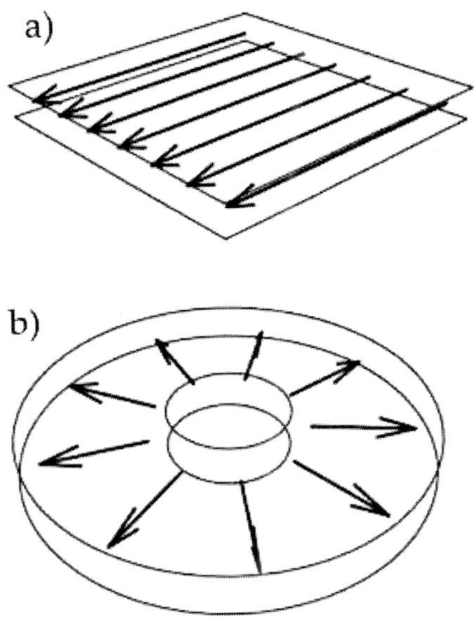

Figure: Visualisation of a) parallel flow and b) radial flow.

$$\frac{d^2u}{dy^2} = -1 \quad ; \quad u(0) = u(1) = 0.$$

The boundary condition is the no slip condition. This problem is easily solved for the flow field:

$$u(y) = \frac{y - y^2}{2}.$$

From this point onward more quantities of interest can be easily obtained, such as viscous drag force or net flow rate.

Difficulties may arise when the problem becomes slightly more complicated. A seemingly modest twist on the parallel flow above would be the *radial* flow between parallel plates; this involves convection and thus nonlinearity. The velocity field may be represented by a function $f(z)$ that must satisfy:

$$\frac{d^2 f}{dz^2} + Rf^2 = -1 \quad ; \quad f(-1) = f(1) = 0.$$

This ordinary differential equation is what is obtained when the Navier–Stokes equations are written and the flow assumptions applied (additionally, the pressure gradient is solved for). The nonlinear term makes this a very difficult problem to solve analytically (a lengthy implicit solution may be found which involves elliptic integrals and roots of cubic polynomials). Issues with the actual existence of solutions arise for R > 1.41 (approximately; this is not the square root of 2), the parameter R being the Reynolds number with appropriately chosen scales. This is an example of flow assumptions losing their applicability, and an example of the difficulty in "high" Reynolds number flows.

Wyld Diagrams

Wyld diagrams are bookkeeping graphs that correspond to the Navier–Stokes equations via a perturbation expansion of the fundamental continuum mechanics. Similar to the Feynman diagrams in quantum field theory, these diagrams are an extension of Keldysh's technique for nonequilibrium processes in fluid dynamics. In other words, these diagrams assign graphs to the (often) turbulent phenomena in turbulent fluids by allowing correlated and interacting fluid particles to obey stochastic processes associated to pseudo-random functions in probability distributions.

Navier–Stokes Equations Use in Games

The Navier–Stokes equations are used extensively in video games in order to model a wide variety of natural phenomena. These include simulations of effects such as water, fire, smoke etc. Many of the implementations used are based on the seminal paper "Real-Time Fluid Dynamics for Games" by J. Stam. More recent implementations based upon this work run on the GPU as opposed to the CPU and achieve a much higher degree of performance.

Reynolds Transport Theorem

Reynolds transport theorem (also known as the Leibniz-Reynold's transport theorem), or in short Reynold's theorem, is a three-dimensional generalisation of the Leibniz integral rule. This theorem is used to compute derivatives of integrated quantities.

Reynolds transport theorem can be simply stated as-What was already there plus what goes in minus what comes out is equal to what is there. Reynolds theorem is used in formulating the basic conservation laws of continuum mechanics, particularly fluid dynamics and large-deformation solid mechanics. These conservation laws (law of conservation of mass, law of conservation of linear momentum, law of conservation of angular momentum, and law of conservation of energy) are adopted from classical mechanics and thermodynamics where the system approach is normally followed. In fluid mechanics, it is often more convenient to work with control volumes as it is difficult to identify and follow a system of fluid particles. Thus, there is a need to relate the system equations and corresponding control volume equations. The link between the two is given by the Reynolds transport theorem. The theorem is named after Osborne Reynolds (1842–1912).

Imagine a system and a coinciding control volume with a control surface. Reynolds transport theorem states that the rate of change of an extensive property *N*, for the system is equal to the time rate of change of *N* within the control volume and the net rate of flux of the property *N* through the control surface. For an example, the law of conservation of mass states that rate of change of the property, mass, is equal to the sum of the rate of accumulation of mass within a control volume and the net rate of flow of mass across the control surface. The differential forms of these equations with additional assumption of Newton's viscosity law are commonly known as the Navier-Stokes equations.

General Form

The Reynolds transport theorem refers to any extensive property, N, of the fluid in a particular control volume. It is expressed in terms of a total derivative on the left-hand side.

$$\frac{dN_{\text{sys}}}{dt} = \int_{\text{c.v.}} \frac{\partial}{\partial t}(\rho\eta)dV + \int_{\text{c.s.}} \rho\eta\vec{\upsilon}_b \cdot \hat{n}dA + \int_{\text{c.s.}} \rho\eta\vec{\upsilon}_r \cdot \hat{n}dA,$$

where η is the intensive property related to extensive property N, i.e. the concentration of N per unit mass; t is time, c.v. refers to the control volume, c.s. refers to the control surface, ρ is the fluid density, V is the volume, υ_b is the velocity of the boundary of the control volume (the control surface), υ_r is the velocity of the fluid relative to the control surface

$$\vec{\upsilon}_r = \vec{\upsilon} - \vec{\upsilon}_b,$$

n is the outward pointing normal vector on the control surface, and A is the area.

Mass Formulation

Also called the continuity equation, the control volume form of the conservation of mass is found by substituting mass in for N. This means that η is equal to 1.

$$0 = \int_{c.v.} \frac{\partial \rho}{\partial t} dV + \int_{c.s.} \rho \vec{v}_b \cdot \hat{n}\, dA + \int_{c.s.} \rho \vec{v}_r \cdot \hat{n}\, dA$$

All variables are defined as in the general formulation. M is equal to the mass of the control volume. Applying the Conservation of mass principle, the left hand side reduces to 0 since mass of a system cannot change in time.

In a steady flow system, the first term on the right hand side of the equation will be equal to 0, i.e. the mass of the control volume does not change, implying that the mass flow rate into the control volume is equal to the mass flow rate out of the control volume.

Momentum Formulation

The momentum equation is found by substituting momentum in for N. From this, η is found to be velocity. From Newton's second law, we have the time rate of change of momentum (now the left hand side of the equation) is equal to the net force. Thus,

$$\sum \vec{F} = \int_{c.v.} \frac{\partial}{\partial t}(\rho \vec{\upsilon}) dV + \int_{c.s.} \rho \vec{\upsilon}(\vec{\upsilon}_b \cdot \hat{n}) dA + \int_{c.s.} \rho \vec{\upsilon}(\vec{\upsilon}_r \cdot \hat{n}) dA,$$

where F is force, c υ is the velocity of fluid in a coordinate system attached to the control surface, and all other variables are defined as in the general formulation. Note that the integral form of the momentum equation is a vector equation.

Energy Formulation

The energy equation is found by substituting energy in for N. From this, η is found to be energy per unit mass.

$$\dot{Q} - \sum \dot{W} = \int_{c.v.} \frac{\partial}{\partial t}\left[\rho\left(\frac{\upsilon^2}{2} + gz + \tilde{u}\right)\right] dV + \int_{c.s.}\left[\frac{\upsilon^2}{2} + gz + \tilde{u} + \frac{p}{\rho}\right]\rho \vec{\upsilon}_b \cdot \hat{n} dA + \int_{c.s.}\left[\frac{\upsilon^2}{2} + gz + \tilde{u} + \frac{p}{\rho}\right]\rho \vec{\upsilon}_r \cdot \hat{n} dA$$

where Q is the heat transfer into the control volume, W is the work done by the system, g is the acceleration due to gravity, z is the vertical distance from an arbitrary datum, υ is the specific internal energy of the fluid, p is the pressure and all other variables are defined as in the general formulation.

Note that these equations make no consideration for chemical reactions or potential energy associated with electromagnetic fields.

Formulation Used in Solid Mechanics

Suppose $\Omega(t)$ is a region in Euclidean space with boundary $\partial\Omega(t)$, and let $\mathbf{n}(\mathbf{x},t)$ be the outward unit normal to the boundary at time t. Let $\mathbf{x}(t)$ be the positions of points in the region, $\mathbf{v}(\mathbf{x},t)$ the velocity field in the region, and let $\mathbf{f}(\mathbf{x},t)$ be a vector field in the region (it may also be a scalar field). Reynolds' transport theorem states that

$$\frac{\mathrm{d}}{\mathrm{d}t}\left(\int_{\Omega(t)} \mathbf{f}\,\mathrm{dV}\right) = \int_{\Omega(t)} \frac{\partial \mathbf{f}}{\partial t}\mathrm{dV} + \int_{\partial\Omega(t)} (\mathbf{v}\cdot\mathbf{n})\mathbf{f}\,\mathrm{dA}\ .$$

A Special Case

If we take Ω to be constant with respect to time, then and $\mathbf{v}_b = 0$ the identity reduces to

$$\frac{\mathrm{d}}{\mathrm{d}t}\int_{\Omega} f\,\mathrm{dV} = \int_{\Omega} \frac{\partial f}{\partial t}\mathrm{dV}\ ,$$

as expected. This simplification is not possible if an incorrect form of the Reynolds transport theorem is used.

Interpretation and Reduction to One Dimension

The theorem is the higher dimensional extension of Differentiation under the integral sign and should reduce to that expression in some cases. Suppose f is independent of y & z, and that $\Omega(t)$ is a unit square in the plane $y-z$ and has x limits $a(t)$ and $b(t)$. Then Reynolds transport theorem reduces to

$$\frac{\mathrm{d}}{\mathrm{d}t}\int_{a(t)}^{b(t)} f\,\mathrm{dx} = \int_{a(t)}^{b(t)} \frac{\partial f}{\partial t}\mathrm{dx} + \frac{\partial b(t)}{\partial t} f(b(t),t) - \frac{\partial a(t)}{\partial t} f(a(t),t)\ ,$$

which is the expression given on Differentiation under the integral sign, except that there the variables x and t have been swapped.

Pipe Networks

(a) Continuity: all flows at a point sum to zero

Notes:

- Flow in is positive, flow out is negative (or vice-versa, but be consistent).
- This assumes no storage of fluid in the system.
- It is valid for steady or unsteady flow.
- In volume flow terms, it applies only to incompressible flows. In mass flow terms, it is general.

(b) Only one pressure can exist at any point. Thus the pressure difference between two points is independent of the route between them.

These rules are similar to those for electrical networks, and can be used similarly to find the "current" through various "resistances" and the "potential" at any point.

Flow will always go from a high pressure "node" to a lower pressure one. The pressure difference will depend on the flow rate, and also on any changes in height or velocity, via Bernoulli's equation. The frictional loss of pressure will appear as a loss of Total Pressure; according to other factors, this may appear as a loss of height, velocity or static pressure.

Pipe Resistance and Reynolds Number Effects

For a given flow rate, the Reynolds number and pipe roughness may be used, with the pipe friction chart, to calculate pressure drop, as in Part I. A copy of the friction chart is in Appendix I and also in the Engineering Data Book. There are also copies in most textbooks, but beware of alternative definition of friction factor f which is four times as big in some books (mainly American, including White).

If the pressure drop is known, and the flow rate required, iteration is needed.

Most pipe flows on a reasonable scale are likely to be turbulent. Follow this procedure:

(i) Estimate a pipe relative roughness

(ii) Assume full turbulence (f independent of Re) and use the pipe friction chart to estimate f.

(iii) Use this to calculate the flow.

(iv) Calculate a new Reynolds number from this flow. Use this to find a new value of f from the friction chart, and iterate back to (iii) until it converges.

If the flow is turbulent, this iteration will usually converge in about 2 loops. If it turns out to be laminar, it may take longer (but the laminar flow relationships-Poiseuille's equation, which is analogous to Ohm's law-can then be used to get an explicit answer). If it seems to be transitional, it may not converge (and there may be no unique correct solution).

An alternative approach is to plot a chart of fRe^2 (or) against Re for various roughness. This can be done from the data on the normal chart (and appears in some elderly textbooks). fRe^2 is independent of velocity, and can be calculated from the pressure drop and constants. Thus Re can be found, whence velocity and everything else. (This effectively uses a slightly different set of dimensionless groups from the normal ones, and is a good example of how the choice of groups can affect the ease of solution.)

Component Loss Coefficient

Assuming turbulent flow, forces tend to vary as square of speed. Define a *component pressure loss coefficient*:

$$k = \frac{\Delta p}{\frac{1}{2}\rho u^2}$$

For many components, this is roughly constant, independent of Re, and dependent only on geometry and installation (later).

A typical value of k is 0.5. 5 is a high value, 0.1 is a low value.

We can obtain a loss coefficient for a straight pipe from the Darcy formula:

$$\Delta p = 4f\frac{l}{d}\frac{1}{2}\rho u^2, \quad k = 4f\frac{l}{d}$$

A typical value of f is 0.005. The loss coefficient for 100 diametres of pipe is therefore 4´0.005´100 = 2.

Sometimes losses are expressed in terms of equivalent length of straight pipe. For the case above, a component with a loss coefficient of 0.5 would have an insertion loss equivalent length of 25 diametres. Note that this depends on the value of f assumed.

If there are not many components in a long length of pipe, their effect will be small compared to friction. They are therefore often called *minor* losses. This does not necessarily imply *negligible.*

Electrical Analogy

In some respects, fluid circuits can be considered analogous to electrical circuits, with pressure replacing voltage and flow rate replacing current.

Then (power) = (pressure drop)´(flow rate) and (resistance) = (pressure drop)/(flow rate)

There are two major differences: firstly, there is no real equivalent in electrical terms of the pressure losses in components (except perhaps for contact resistance), and secondly, there is no such thing as electrical turbulence. Resistance, as defined above, is a constant for a given pipe only in laminar flow. In turbulent flow, (pressure drop)/(flow rate)^2 is more nearly constant, which is the reason for the definition of k above. The analogy can be extended to circuits in series or in parallel.

Two systems in series have the same flow rate. The total pressure drop is the sum of the individual pressure drops. Assuming constant velocity (and therefore cross-sectional area),

$$k_{total} = \sum k_i$$

This is identical to the resistors in series formula. (In the event of varying velocity, some convention as to the velocity to be used in defining k must be adopted, and the cross-sectional areas will appear in the above equation).

Two systems in parallel have the same pressure drop. The flow rate distributes to ensure this:

$$Q = \varepsilon Q_i = \varepsilon A_i u_i$$

$$\Delta p = \tfrac{1}{2} \rho u_i^2 k_i$$

For constant cross-sectional areas, these equations can be combined to give:

$$\frac{1}{\sqrt{k_{total}}} = \sum \frac{1}{\sqrt{k_i}}$$

This is similar to the resistors in parallel formula, but less useful because of the effects of cross-sectional area change, and the square roots.

Branching Networks and Rings

Some networks have branches and loops, and it may not be obvious which way the flow goes in some elements. Since the pressure falls in the flow direction, we need to know direction to calculate pressure drop, and vice-versa. In laminar flow (as in electrical networks), we can just make a guess, and it will come out negative if we were wrong. This is not the case with turbulent flow, since the sign gets lost in squaring. In principle, all possible combinations of flow directions must be tested, and those producing physically unrealistic results discarded (e.g. negative pressures, or failing due to square roots of negative numbers). Except for the simplest networks, this is impractical.

Another approach is to assume the flow in such an element is zero, and solve the resulting network. This will then suggest which way the flow is going (which end is the pressure higher?), and thus enable the full problem to be solved. This will usually work. If it doesn't, various iterative methods (and computer programs) are available.

Interactions

If two components (e.g. bends) are close together, they may interact, and the overall pressure loss may not be the sum of the two separately:

- Exit conditions of first modified by presence of second. Some of the separation and turbulence losses happen in the pipe after the component;
- Entry conditions of second affected by presence of first: turbulence, velocity profile, secondary flow.

$$k_{total} = C(k_1 + k_2)$$

C = interaction correction factor.

Spacing:

>30 diametres: $C = 1$ (no effect)

>3 diametres: $C < 1$ usually (beneficial)

<3 diametres: $0.5 < C < 1.1$ usually

Changes of Section

Expansion and Exit

For a sudden or abrupt expansion, $k = \left(1 - \frac{A_1}{A_2}\right)^2$, based on the upstream (higher) velocity.

For an exit, $A_2 = \infty$ and $k = 1$. This represents a loss of the dynamic pressure of the flow. This so-called exit loss is not really a loss-the energy has not been degraded to heat and could (in principle) be recovered. Rather, it is kinetic energy leaving the system.

A gradual expansion is called a diffuser. Depending on the area ratio and the angle (assuming the diffuser is conical), there may or may not be separation losses. Diffusers form a vital part of many power systems (jet engines, venturis, water turbines, etc). Efficient diffuser design is a specialist field.

There is no general theory, so we use the sudden expansion theory with a diffuser coefficient K:

$$k = K\left(1 - \frac{A_1}{A_2}\right)^2$$

K depends on the area ratio, the cone angle and the Reynolds number, in no simple manner. The lowest value is about 0.15 for an angle of 6°, so that 6° diffusers are often considered standard. The worst case is about 1.25 at an angle of about 60°. Thus a sudden expansion (90°) may be better than an attempt at a short diffuser.

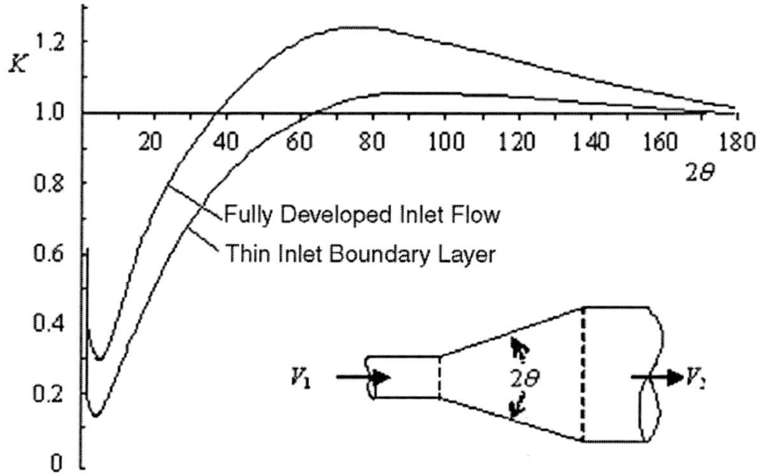

Contraction and Entry

For a sudden or abrupt contraction, the loss coefficient is given by $k = \left(\frac{1}{C_C} - 1\right)^2$, based on the higher (outlet) velocity, and C_C is the contraction coefficient which depends on area ratio, but is normally around 0.6. (This result is derived in many text books).

A gradual contraction with smoothly curved sides can be designed to avoid separation and thus have very low loss-no more than the friction component. $k < 0.1$. The general approach is that the curve should be continuous in position (no gaps), slope (no corners) and curvature (one curve blending into the next smoothly).

For an entry, the inlet area is infinite. The loss coefficient is very dependent on the detail geometry of the entry. Values range from $k = 1$ for a long spigot protruding into the reservoir to $k < 0.1$ for a rounded, faired or bell entry. Even a simple bevel (countersunk) entry has quite a low value-around 0.2.

Fittings and Components

Bends have a k value between about 0.1 and 1.1, depending on geometry.

Filters, Screens, Gauzes, Grids

Filters: pore size usually so small that flow through filter is laminar. Dp μ Q. (k not a useful quantity, since it varies with flow rate).

Gauzes etc: define *solidity* as

$$S = \frac{\text{Closed Area}}{\text{Total Area}}$$

For example, consider a square mesh, with cell size h and wire diametre d. The total area of a cell is h^2, while the open area is $(h-d)^2$. The solidity is therefore

$$S = \frac{h^2 - (h-d)^2}{h^2} = \frac{d}{h}\left(2 - \frac{d}{h}\right)$$

(The same formula turns out to be valid for triangular and hexagonal meshes, and similar ones can be derived for other shapes.)

Experiments relate k to S. For example, at wire Reynolds numbers Re_d above about 300, it is found that $k = 0.9\frac{S}{(1-S)^2}$.

Valves

The purpose of a valve is to produce a pressure drop. A closed (non-leaking) valve has a k of ∞. A fully open valve has a minimum value of k which depends on its design, ranging from < 0.1 to 10 or more. A partly open valve produces a variation of k with control position, which varies with valve type. For control purposes, a smoothly varying characteristic is best. For shut-off or isolation, a sudden change may be better.

Flow Metres

Pressure drop type typically k = 0.5 for venturimetre, k = 8 for orifice plate.

Turbine type: typically k = 1.

Positive displacement types (petrol pump, gas metre) and "float" (Rotametre) types: essentially fixed force, and therefore fixed pressure drop; k falls with flow rate. Typically k = 10 at low flow, k = 1 at full flow, but could be much higher.

Laminar flow

Laminar flow, sometimes known as streamline flow, occurs when a fluid flows in parallel layers, with no disruption between the layers.

At low velocities the fluid tends to flow without lateral mixing, and adjacent layers slide past one another like playing cards. There are no cross currents perpendicular to the direction of flow, nor eddies or swirls of fluids. In laminar flow the motion of the particles of fluid is very orderly with all particles moving in straight lines parallel to the pipe walls.

In fluid dynamics, laminar flow is a flow regime characterised by high momentum diffusion and low momentum convection.

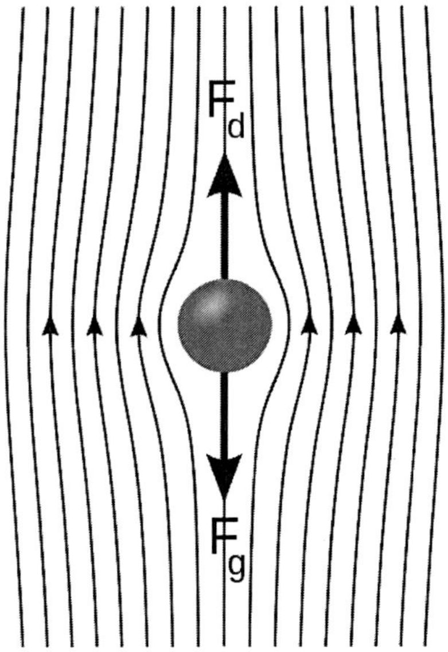

Figure: An object moving through a gas or liquid experiences a force in direction opposite to its motion. Terminal velocity is achieved when the drag force is equal in magnitude but opposite in direction to the force propelling the object. Shown is a sphere in Stokes flow, at very low Reynolds number.

When a fluid is flowing through a closed channel such as a pipe or between two flat plates, either of two types of flow may occur depending on the velocity of the fluid: laminar flow or turbulent flow. Laminar flow is the opposite of turbulent flow which occurs at higher velocities where eddies or small packets of fluid particles form leading to lateral mixing. In nonscientific terms laminar flow is "smooth", while turbulent flow is "rough."

The type of flow occurring in a fluid in a channel is important in fluid dynamics problems. The dimensionless Reynolds number is an important parameter in the equations that describe whether flow conditions lead to laminar or turbulent flow. In the case of flow through a straight pipe with a circular cross-section, at a Reynolds number below the critical value of approximately 2040 fluid motion will ultimately be laminar, whereas at larger Reynolds number the flow can be turbulent. The Reynolds number delimiting laminar and turbulent flow depends on the particular flow geometry, and moreover, the transition from laminar flow to turbulence can be sensitive to disturbance levels and imperfections present in a given configuration.

When the Reynolds number is much less than 1, Creeping motion or Stokes flow occurs. This is an extreme case of laminar flow where viscous (friction) effects are much greater than inertial forces. The common application of laminar flow would be in the smooth flow of a viscous liquid through a tube or pipe. In that case, the velocity of flow varies from zero at the walls to a maximum along the centreline of the vessel. The flow profile of laminar flow in a tube can be calculated by dividing the flow into thin cylindrical elements and applying the viscous force to them.

For example, consider the flow of air over an aircraft wing. The boundary layer is a very thin sheet of air lying over the surface of the wing (and all other surfaces of the aircraft). Because air has viscosity, this layer of air tends to adhere to the wing. As the wing moves forward through the air, the boundary layer at first flows smoothly over the streamlined shape of the airfoil. Here the flow is called *laminar* and the boundary layer is a laminar layer. Prandtl applied the concept of the laminar boundary layer to airfoils in 1904.

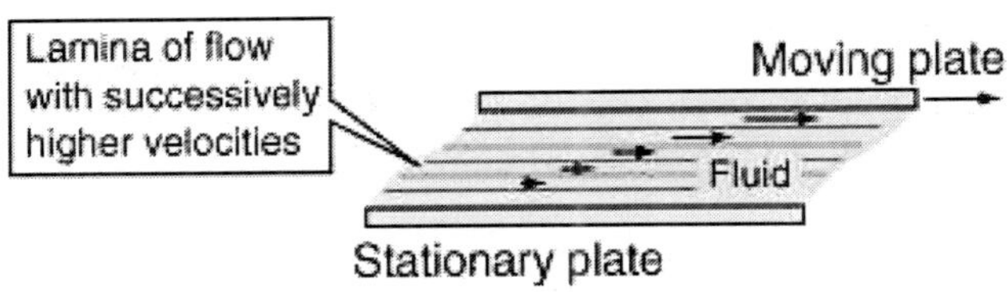

Figure: In the case of a moving plate in a liquid, it is found that there is a layer or lamina which moves with the plate, and a layer which is essentially stationary if it is next to a stationary plate.

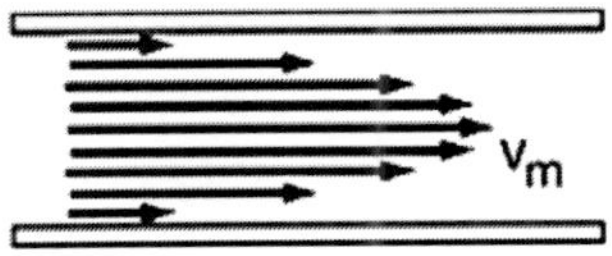

Figure: The streamlines associated with laminar flow resemble a deck of cards. This flow profile of a fluid in a pipe shows that the fluid acts in layers and slides over one another.

Besides, laminar flow reactor (LFR) is a reactor that uses laminar flow to study chemical reactions and process mechanisms.

Chapter 6

Poiseuille's Law and the Viscosity of Fluids

The simplest model for the flow of a fluid through a pipe is contained in Bernoulli's equation, which is simply an application of the law of conservation of mechanical energy to a moving fluid:

$$P + pgh + \frac{pv^2}{2} = cons\tan t$$

here P is the pressure of the fluid, r its density, h its height and v its velocity. This model assumes that the fluid is incompressible and that it suffers no friction as it moves through the pipe. The use of a single value for v in Bernoulli's law indicates that it is based on the assumption that every portion of the fluid moves with the same velocity. In reality that is clearly not the case: the part of the fluid that is in direct contact with the wall will move very slowly while elsewhere the velocity will be higher. The internal friction that causes this velocity gradient is called the viscosity of the fluid.

For sufficiently small velocities the flow will be laminar i.e. layered. One can show that laminar flow leads to a parabolic distribution of velocities across a pipe with a circular cylindrical cross section. As the velocity increases past a critical value, depending upon the viscosity and density of the fluid, eddies appear and the flow becomes turbulent.

The laminar flow through a pipe is described by the Hagen-Poiseuille law, stating that the flow rate (F = volume of fluid flowing per unit time) is proportional to the pressure difference Dp between the ends of the pipe and the fourth power of its radius r.

$$F = \frac{\pi pr^4}{8\mu 1}$$

Here is the length of the pipe and m the coefficient of viscosity, a constant characteristic of the given fluid. It should be noted that, for liquids the coefficient of viscosity decreases, i.e., the liquid becomes less viscous, with increasing temperature. (We make here a distinction between a fluid, any medium that can flow, including gases, and liquids such as water. Experimentally, one can measure the coefficient of viscosity of a given liquid by observing its flow rate through a pipe of a given diametre under the effect of a known pressure. Obviously, the measurement should be performed while dealing with laminar, not turbulent flow: the presence of such a condition can be verified by observing the proportionality between flow rate and applied pressure.

The onset of turbulent flow is indicated by a departure from the linear beahviour.

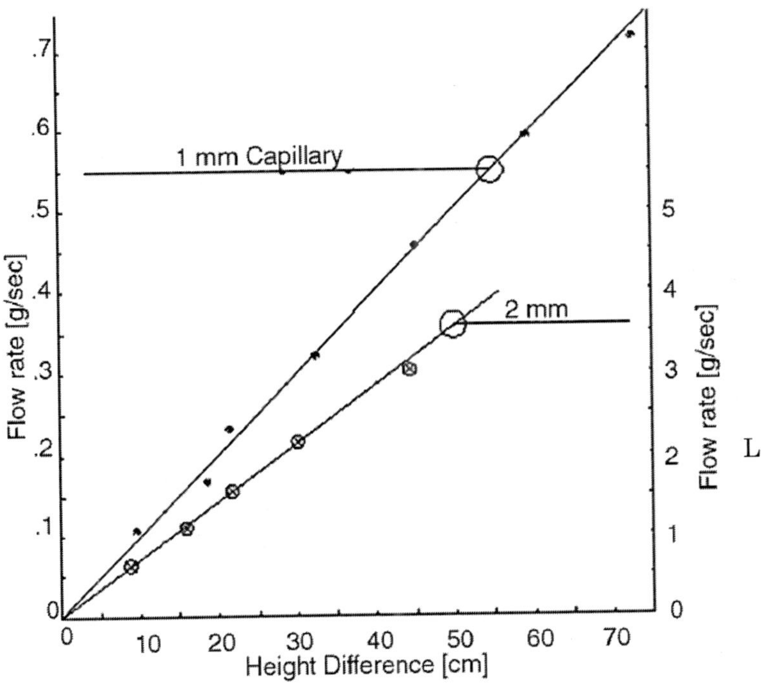

Figure: Linear dependence of flow rate.

Note the strong (fourth power) dependence of the flow rate on the pipe radius: even a small variation of the radius will cause a large variation of the flow rate. Ergo, don't be too disturbed by deviations from the r^4 law that you might observe: a 5% deviation from the given radius of the pipe will lead to a 20% change in the flow rate.

Because the flowrate F and the pressure p are proportional to each other (Eq. 2)-as are current and voltage in Ohm's law-it is convenient to write Eq. 2 in a form analogous to Ohm's law by introducing the flow resistance R, defined as:

$$R = \frac{p}{F} = \frac{8\pi L}{\pi r^4}$$

Apparatus

Water bottle; bottle stand; plastic tube; clamp; set of 3 capillary tubes of different diametres; various adapters; stand and clamps for capillaries; basin; beaker; funnel; metre stick; spirit level; timer; scales; lab jack..

What to do

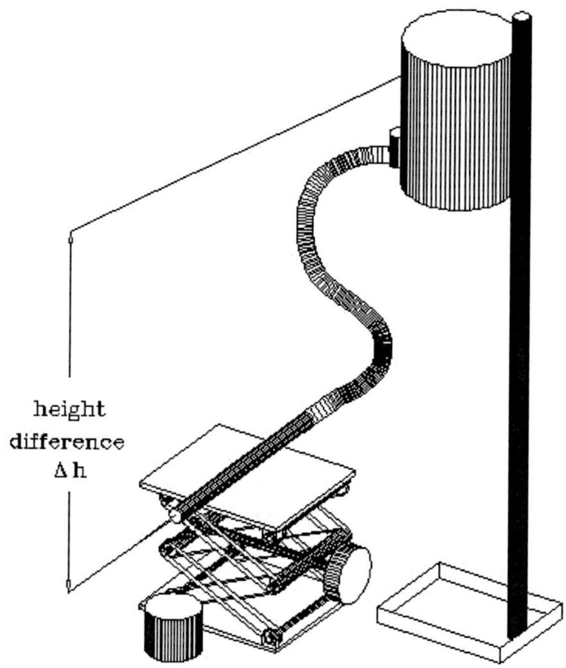

Figure: shows a schematic of the apparatus.

- Insert the 2 mm capillary into the end of the plastic tube and tape it to the lab jack with masking tape. Make sure the capillary is level, i.e. don't let the plastic tube rest on the jack. Fill the bottle with water and mount it so that the fluid level is 5 to 10 cm above the capillary. The height difference between the two determines the pressure difference

$$\Delta p = p\Delta g$$

which, in turn, determines the flow rate.

- Dry the plastic beaker and the pans of the scale. Weigh the empty beaker.
- Start the water flow, making sure that the stopcock (the hose clamp) is fully open and that there are no air bubbles moving through the capillary. Holding the stop watch in one hand and the dry beaker in the other, insert the latter into the water stream while starting the former. Collect some 20 to 40 cm^3 of water and withdraw the beaker while stopping the watch. Then, immediately, stop the water flow by shutting the stopcock (the hose clamp).
- Measure the height difference, Dh, between the water level in the bottle and the axis of the capillary.
- Weigh the amount of water that you have collected and calculate the flow rate F, taking the density of water to be 1 g/cm^3.
- Pour the collected amount of water back into the bottle and measure the change in the water level. If necessary make a correction by extrapolating to the level difference during the middle of the flow rate measurement.
- Plot the flow rate as a function of the height difference.
- Dry the beaker and the scale.
- Raise the bottle to a new level and go back to step 3.
- Repeat steps 3) to 9) until you have reached the full height of the stand or until the last point shows a clear deviation from a linear depency on the pressure difference.
- Repeat the entire procedure first for the 1.5 and then for the 1 mm capillary. As the diametre decreases, it may be necessary to start the measurements at a larger initial value of the height difference and, for the smallest pipe, to collect the water for up to two minutes.

In your lab report determine for each capillary the value of m. Are the fluctuations of the individual values consistent with the accuracy of the measurements? Calculate for each capillary the flow resistance $R = p/F$, and plot it versus r^{-4}; according to Poiseuille's law you should get a linear dependence, do you?

In actuality you measure the flow resistance of the capillary in series with the plastic tubing. Is this an effect to be taken into account? How would this effect manifest itself? How can you correct for it?

Power-law Fluid

A Power-law fluid, or the Ostwald–de Waele relationship, is a type of generalised Newtonian fluid for which the shear stress, τ, is given by

$$\tau = K\left(\frac{\partial u}{\partial y}\right)^{n}$$

where:

- K is the *flow consistency index* (SI units Pa•s^n),
- "u/"y is the shear rate or the velocity gradient perpendicular to the plane of shear (SI unit s^{-1}), and
- n is the *flow behaviour index* (dimensionless).

The quantity

$$\mu_{\text{eff}} = K\left(\frac{\partial u}{\partial y}\right)^{n-1}$$

represents an *apparent* or *effective viscosity* as a function of the shear rate (SI unit Pa•s).

Also known as the Ostwald–de Waele power law this mathematical relationship is useful because of its simplicity, but only approximately describes the behaviour of a real non-Newtonian fluid. For example, if n were less than one, the power law predicts that the effective viscosity would decrease with increasing shear rate indefinitely, requiring a fluid with infinite viscosity at rest and zero viscosity as the shear rate approaches infinity, but a real fluid has both a minimum and a maximum effective viscosity that depend on the physical chemistry at the molecular level. Therefore, the power law is only a good description of fluid behaviour across the range of shear rates to which the coefficients were fitted. There are a number of other models that better describe the entire flow behaviour of shear-dependent fluids, but they do so at the expense of simplicity, so the power law is still used to describe fluid behaviour, permit mathematical predictions, and correlate experimental data.

Power-law fluids can be subdivided into three different types of fluids based on the value of their flow behaviour index:

Pseudoplastic Fluids

Pseudoplastic, or shear-thinning fluids have a lower apparent viscosity at higher shear rates, and are usually solutions of large, polymeric molecules in a solvent with smaller molecules. It is generally supposed that the large molecular chains tumble at random and affect large volumes of fluid under low shear, but that they gradually align themselves in the direction of increasing shear and produce less resistance.

A common household example of a strongly shear-thinning fluid is styling gel, which primarily composed of water and a fixative such as a vinyl acetate/vinylpyrrolidone copolymer (PVP/PA). If one were to hold a sample of hair gel in one hand and a sample of corn syrup or glycerine in the other, they would find that the hair gel is much harder to pour off the fingers (a low shear application), but that it produces much less resistance when rubbed between the fingers (a high shear application).

Newtonian Fluids

A Newtonian fluid is a power-law fluid with a behaviour index of 1, where the shear stress is directly proportional to the shear rate:

$$\tau = \mu \frac{\partial u}{\partial y}$$

These fluids have a constant viscosity, μ, across all shear rates and include many of the most common fluids, such as water, most aqueous solutions, oils, corn syrup, glycerine, air and other gases.

While this holds true for relatively low shear rates, at high rates most oils in reality also behave in a non-Newtonian fashion and thin. Typical examples include oil films in automotive engine shell bearings and to a lesser extent in geartooth contacts.

Dilatant Fluids

Dilatant, or shear-thickening fluids increase in apparent viscosity at higher shear rates. They are rarely encountered, but one common example is an uncooked paste of cornstarch and water. Under high shear the water is squeezed out from between the starch molecules, which are able to interact more strongly.

While not strictly a dilatant fluid, Silly Putty is an example of a material that shares these viscosity characteristics. Another use is in a viscous coupling in which if both ends of the coupling are spinning at the same (rotational) speed, the fluid viscosity is minimal, but if the ends of the coupling differ greatly in speed, the coupling fluid becomes very viscous. Such couplings have applications as a lightweight, passive mechanism for a passenger automobile to automatically switch from two-wheel drive to four-wheel drive such as when the vehicle is stuck in snow and the primary driven axle starts to spin due to loss of traction under one or both tires.

Velocity Profile in a Circular Pipe

Just like a Newtonian fluid in a circular pipe gives a quadratic velocity profile, a power-law fluid will result in a power-law velocity profile,

$$u(r) = \frac{n}{n+1}\left(\frac{dp}{dz}\frac{1}{2K}\right)^{\frac{1}{n}}\left[R^{\frac{n+1}{n}} - r^{\frac{n+1}{n}}\right]$$

where $u(r)$ is the (radially) local axial velocity, dp/dz is the pressure gradient along the pipe, and R is the pipe radius.

Newtonian Fluid

A Newtonian fluid (named after Isaac Newton) is a fluid whose stress versus strain rate curve is linear and passes through the origin. The constant of proportionality is known as the viscosity.

Definition

A simple equation to describe Newtonian fluid beahviour is

$$\tau = \mu\frac{du}{dy}$$

where

- tis the shear stress exerted by the fluid ("drag") [Pa]
- μmis the fluid viscosity-a constant of proportionality [Pa ·s]
- $\frac{du}{dy}$ is the velocity gradient perpendicular to the direction of shear, or equivalently the strain rate [s^{-1}]

In common terms, this means the fluid continues to flow, regardless of the forces acting on it. For example, water is Newtonian, because it continues to exemplify fluid properties no matter how fast it is stirred or mixed.

Other examples may be aqueous solutions, emulsions. Contrast this with a non-Newtonian fluid, in which stirring can either leave a "hole" behind (that gradually fills up over time-this beahviour is seen in materials such as pudding and oobleck, or, to a less rigorous extent, sand), or climb the stirring rod (the Weissenberg effect) because of shear thinning, the drop in viscosity causing it to flow more (this is seen in non-drip paints, which brush on easily but become more viscous when on walls).

For a Newtonian fluid, the viscosity, by definition, depends only on temperature and pressure (and also the chemical composition of the fluid if the fluid is not a pure substance), not on the forces acting upon it. If the fluid is incompressible and viscosity is constant across the fluid, the equation governing the shear stress, in the Cartesian coordinate system, is the tensor

$$\tau_{ij} = \mu\left(\frac{\partial u_i}{\partial x_j} + \frac{\partial u_j}{\partial x_i}\right)$$

A second tensor, P (also written as σ), representing the total stress, can be written by combining the shear stress with conventional (thermodynamic) pressure:

$$\mathbb{P}_{ij} = -p\delta_{ij} + \mu\left(\frac{\partial u_i}{\partial x_j} + \frac{\partial u_j}{\partial x_i}\right)$$

where, by the convention of tensor notation,

- τ_{ij} is the shear stress on the i th face of a fluid element in the j th direction
- $P$is thermodynamical pressure
- u_i is the velocity in the i th direction
- is the j th direction coordinate

If a fluid does not obey this relation, it is termed a non-Newtonian fluid, of which there are several types, including polymer solutions, molten polymers, many solid suspensions and most highly viscous fluids.

Non-Newtonian Fluid

A non-Newtonian fluid is a fluid whose flow properties differ in any way from those of Newtonian fluids. Most commonly the viscosity (resistance to deformation or other forces) of non-Newtonian fluids is dependant on shear rate or shear rate history. However, there are some non-Newtonian fluids with shear-independent viscosity, that nonetheless exhibit normal stress-differences or other non-Newtonian behaviour. Many salt solutions and molten polymers are non-Newtonian fluids, as are many commonly found substances such as ketchup, custard, toothpaste, starch suspensions, paint, blood, and shampoo.

In a Newtonian fluid, the relation between the shear stress and the shear rate is linear, passing through the origin, the constant of proportionality being the coefficient of viscosity. In a non-Newtonian fluid, the relation between the shear stress and the shear rate is different, and can even be time-dependent.

Therefore, although the concept of viscosity is commonly used in fluid mechanics to characterise the shear properties of a fluid, it can be inadequate to describe non-Newtonian fluids. They are best studied through several other rheological properties which relate stress and strain rate tensors under many different flow conditions, such as oscillatory shear, or extensional flow which are measured using different devices or rheometres. The properties are better studied using tensor-valued constitutive equations, which are common in the field of continuum mechanics.

Shear Thinning Fluid

A familiar example of the opposite, a shear thinning fluid, or pseudoplastic fluid, is paint: one wants the paint to flow readily off the brush when it is being applied to the surface being painted, but not to drip excessively.

Bingham Plastic

There are fluids which have a linear shear stress/shear strain relationship which require a finite yield stress before they begin to flow (the plot of shear stress against shear strain does not pass through the origin). These fluids are called Bingham plastics. Several examples are clay suspensions, drilling mud, toothpaste, mayonnaise, chocolate, and mustard. The surface of a Bingham plastic can hold peaks when it is still. By contrast Newtonian fluids have flat featureless surfaces when still.

Rheopectic

There are also fluids whose strain rate is a function of time. Fluids that require a gradually increasing shear stress to maintain a constant strain rate are referred to as rheopectic. An opposite case of this, is a fluid that thins out with time and requires a decreasing stress to maintain a constant strain rate (thixotropic).

Examples

Oobleck

An inexpensive, non-toxic example of a non-Newtonian fluid is a suspension of starch (e.g. cornstarch) in water, sometimes called "oobleck" or "ooze" (1 part of water to 1.5–2 parts of corn starch). Uncooked imitation custard, being a suspension of primarily cornflour, has the same properties. The name "oobleck" is derived from the children's book *Bartholomew and the Oobleck.*

Figure: Oobleck on a subwoofer. Applying force to oobleck, by sound waves in this case, makes the non-Newtonian fluid thicken.

Flubber

Flubber is a non-Newtonian fluid, easily made from polyvinyl alcohol–based glues and borax, that flows under low stresses but breaks under higher stresses and pressures. This combination of fluid-like and solid-like properties makes it a Maxwell solid. Its behaviour can also be described as being viscoplastic or gelatinous.

Chilled Caramel Topping

Another example of this is chilled caramel ice cream topping. The sudden application of force—for example by stabbing the surface with a finger, or rapidly inverting the container holding it—leads to the fluid behaving like a solid rather than a liquid. This is the "shear thickening" property of this non-Newtonian fluid. More gentle treatment, such as slowly inserting a spoon, will leave it in its liquid state.

Trying to jerk the spoon back out again, however, will trigger the return of the temporary solid state. A person moving quickly and applying sufficient force with their feet can walk across such a liquid.

Silly Putty

Silly Putty is a silicone polymer based suspension which will flow, bounce, or break depending on strain rate.

Ketchup

Ketchup is a shear thinning fluid. Shear thinning means that the fluid viscosity decreases with increasing shear stress. In other words, fluid motion is initially difficult at slow rates of deformation, but will flow more freely at high rates.

Design Equations for Laminar and Turbulent Flow in Pipes

Standard sizes of pipes:

- An important application of fluid flow: flow inside circular conduits, pipes, and tubes.
- Sizes of commercial standard steel pipe are given in A.5.
- There are two main schedule numbers; schedule 40 and schedule 80.
- Schedule 80 has a thicker wall and will hold about twice the pressure of schedule 40.
- Both have the same outside diametre so they will fit the same fittings.
- Schedule 40 is usually used.

Pipelines

Pipelines are the most common means of transporting oil or gas. A pipeline is like any other flowline. The main differences are that pipelines are long and continuously welded, they have a minimum number of curves, they have no sharp bends, and they are most often either buried or otherwise inaccessible due to their location over the majority of their length.

These differences mean that small sections of pipeline are not easily removed for maintenance and consequently great care is taken to prevent problems arising in the first place. A pipeline is extremely expensive to lay, and in the case of offshore pipelines, costs in the order of several million pounds per subsea mile have been encountered.

Maintenance on pipelines is also expensive but this expenditure is necessary since, regardless of the expense, pipelines frequently form the most efficient and cost-effective method of transporting the quantifies of oil or gas produced. Pipeline sharing agreements may result in the flow from a number of oil fields being transported through a single pipeline. A problem in a pipeline of this type can mean the shut-down of all of these fields with a resulting operating loss of several million pounds per day.

This situation can be further aggravated for gas production to gas consumer companies where the producing company can not only lose operating revenue but can incur fines for failing to fulfill contractual obligations.

Pipeline Design

When designing a pipeline, the engineer considers the following factors:

- The physical and chemical properties of the fluid, or to pumped through the pipeline;
- The maximum volume of fluid that will be pumped through the pipeline at any time during the life of current and future developments likely to be served by the pipeline.
- The nature of the environment through which the pipeline is going to traverse.
- The required delivery pressure.

More specifically the engineer considers

- Pipe diametre required. (The larger the diametre of the pipeline, the more fluid can be moved through it, assuming other variables such as pump capacity are fixed.)
- Pipe length. (The greater the length of a segment of pipeline, the greater the total pressure drop. Pressure drop can be the same per unit of length for a given size and type of pipe but total pressure drop increases with length.)
- Specific gravity and density of the fluid to be transported, (The specific gravity and density of the transported fluid will affect the potential amount of mass flow available.)
- Compressibility. (Because most liquids are only slightly compressible, this term is not usually significant in calculating liquids pipeline capacity at normal operating conditions. In gas and gas liquids (mixtures of methane, ethane, propane, butane, etc, transported as a liquid) pipeline design, however, it is necessary to include a term in many design calculations to account for the fact that gases deviate from laws describing ideal gas behaviour under conditions other than standard or base conditions. This term, supercompressibility factor, is very significant at high temperatures and pressures. If in the pipeline, pressure is likely to be in the order of 1000 to 2000 psig then this term must be included.)

- Operating temperatures and ambient temperatures: (Temperature affects pipeline capacity both directly and indirectly. In natural gas pipelines, the lower the operating temperature, the greater the capacity, assuming all other variables are fixed.

Operating temperature also can affect other terms in equations used to calculate the capacity of both liquids and natural gas pipelines. Viscosity, for example, varies with temperature. Designing a pipeline for heavy (viscous) crude is one case in which it is necessary to know operating temperatures accurately to calculate pipeline capacity. The possibility of water freezing and of hydrate formation in gas pipelines are other temperature considerations.

- *Viscosity:* (The property of a fluid that resists flow or relative motion between adjacent parts of the fluid is viscosity. It is an important term in calculating line size and horsepower requirement when designing liquid pipelines).
- *Pour Point:* (The lowest temperature at which an oil will pour, or flow, when cooled under specific test conditions is the pour point. oils can be pumped below their pour points, but the design and operation of a pipeline under these conditions presents special problems.)
- Vapour Pressure. (The pressure that holds a volatile liquid in equilibrium with its vapour at a given temperature is its vapour pressure; when page 73 determined for petroleum products under specific test conditions and using specific procedures it is called the RVP (Reid Vapour Pressure). Vapour pressure is an especially important design criterion when handling volatile petroleum products such as propane or butane.

The minimum pressure in the pipeline must be high enough to maintain these fluids in their liquid state.

Reynolds Number, which is a dimensionless number, which is used to describe the type of flow exhibited by a flowing fluid. In streamlined (or laminar) flow, the molecules move parallel to the axis of flow. In turbulent flow, the molecules move back and forward across the flow axis.

Other types of flow are also possible and the Reynolds number can be used to determine which types of flow are likely to occur under specified conditions. In turn, the type of flow exhibited by a fluid affects pressure drop in the pipeline. Strictly speaking. a Reynolds Number below 1000 describes streamlined flow.

At Reynolds Numbers between 1000 and 2000 flow is unstable. At Reynolds Numbers greater than 2000 flow is turbulent. In general usage, how is considered laminar for $R<2000,>4000$.

Friction Factor. (A variety of friction factors are used in pipeline calculations. They are determined empirically and are related to the roughness of the inside pipe wall)

This is not a complete list but represents the basic parameters used. Terms are interdependent; for example, operation pressure depends on pressure drop, which depends on flow rate, which in turn is dictated by allowable pressure drop.

Several pressure terms are used in pipeline design and operation. Barometric pressure is the value of the atmospheric pressure above a perfect vacuum. A perfect vacuum cannot exist on the earth, but it makes a convenient reference point for pressure measurement.

Absolute pressure is the pressure of a pipeline or vessel above a perfect vacuum and is abbreviated bara. Gauge pressure is the pressure measured in a pipeline or vessel above atmospheric pressure and is abbreviated barg. Standard atmospheric pressure is usually considered to be the head pressure of 760 mm of mercury, but atmospheric pressure varies with elevation above sea level. Many contracts for the purchase of natural gas, for instance, specify that the standard, or base, pressure will be other than 760mm/kg.

Formulas describing the flow of fluids in a pipe are derived from Bernoulli's theorem and are modified to account for losses due to friction. Bernoulli's theorem expresses the application of the law of conservation of energy to the flow of fluids in a conduit To describe the actual flow of gases and liquids properly, however, solutions based on Bernoulli's theorem require the use of coefficients that must be determined experimentally.

As a basic rule, the amount of flow along a pipeline (or across any restriction) will be a function of the differential pressure. The basic equation is:

$$\%Q = \sqrt{\%}\ Dp \times 10$$

where:

0= Flow (In %)

Dp = Differential pressure (in %)

The theoretical equation for fluid flow neglects friction and assumes no energy is added to the systems by pumps or compressors. Of course, in the design and operation of a pipeline, friction losses are very important, and pumps and compressors are required to overcome those losses. So practical pipeline design equations depend on empirical coefficients that have been determined during years of research and testing.

The basic theory of fluid flow does not change. But modifications continue to be made in coefficient as more information is available, and the application of various forms of basic formulas continues to be refined. The use of computers for solving pipeline design problems has also enhanced the accuracy and inflexibility possible in pipeline design.

Liquids Pipelines

In the design of liquids and natural gas pipelines, pressure drop, flow capacity and pumping or compression horsepower required are key calculations. The design of a liquids pipeline is similar in concept to the design of a natural gas pipeline. In both cases, a delivery pressure and the volume the pipeline must handle are known. The allowable working pressure of the pipe can be determined using the pipe size and type and specified safety factors.

In most pipeline calculations, assumptions must be made initially. For instance, a line size may be assumed in order to determine maximum operating pressure and the pressure drop in a given length of pipe for a given flow volume. If the resulting pressure drop, when added to the known delivery pressure exceeds the allowable working pressure, a larger pipe size must usually be chosen.

It may be possible to change the capacity and spacing of booster pumping stations to stay within operating pressure. But in the simplest case, if the calculation yields an operating pressure greater than allowed, a larger pipe size must be selected and the calculation repeated.

It is apparent that many options are available in even a moderately complex pipeline system. But today's computer programs for pipeline design can analyse many variables and many options in a short time, greatly easing the design process.

Pressure Drop

An equation for the flow of liquids in a pipe was developed by Darcy in the early 18th Century and the equations, formulae and standards defined by Darçy are still valid today.

The Darcy equation can be derived mathematically (except for a friction factor which must be determined by experiment) and can be used to calculate for laminar and turbulent flow of liquid in a pipe.

Valves and Fittings

In addition to the pressure loss due to fluid friction with the walls of the pipeline, valves and fittings also contribute to overall system pressure loss. The pressure loss due to a single valve in several thousand feet of straight piping will be insignificant but in a pumping station, for example, where many valves exist and many changes in flow direction occur, pressure loss in valves and fittings is important Pressure loss in valves and fittings is made up of both the friction loss within the valve or fitting itself and the additional loss upstream and downstream of the fitting above that which would have occurred in the absence of the fitting. Calculation of the pressure loss in a valve or fitting is based on experimental data. One approach is the use of a resistance factor for a given valve or fitting. The resistance coefficient is normally treated as a constant for a given valve or fitting under all flow conditions.

Another term used in determining the pressure drop through valves and fittings is the flow coefficient, Cv The flow coefficient of a valve is the flow of water at 6O°F, in gal/min, at a pressure drop of one psi across the valve. The flow coefficients of any other liquid can be calculated using the relation of its density to that of water.

Heavy Crudes

Some crudes with very high pour points or high wax contents that require pipelines of special design Pipelining such crudes can be especially troublesome offshore where heat loss to the water is great and any heat added to the crude before it enters the pipeline is dissipated within a short distance

if a conventional pipeline is used. If the crude cools, excessive wax deposits in the pipeline can lower operating efficiency. In cases of extremely viscous crudes, flow can even be halted if the temperature is allowed to fail too low. Not only is the baiting of flow a problem, but restarting flow after such an occurrence can be difficult. To handle these special crudes, pipelines have been successfully installed and operated simply by insulating the pipelines, but other approaches include:

- Heating the crude to a high temperature at the inlet to the pipeline, allowing it to reach its n destination before cooling below the pour point (The pipeline may or may not be insulated);
- Pumping the crude at a temperature below the pour point using high pressure pumps;

- Adding a hydrocarbon dilutant such as a less waxy crude or a light distillate;
- Injecting water to form a layer between the pipe wall and the crude;
- Processing the crude before pipelining to change the wax crystal structure and reduce pour point and viscosity.
- Mixing water with the crude to form an emulsion; Processing the crude before pipelining to change the wax crystal structure and reduce pour point and viscosity;
- Heating both crude and pipeline by steam tracing or electrical heating;
- Injecting wax solvents such as benzene or toluene.

A combination of these methods can also be used and the choice of method will depend upon the physical properties of the crude and the economics of its production.

If waxy crude is pumped below its pour point, more pumping energy is required and, if pumping is stopped, more energy will be required to put the crude in motion again than was required to keep it flowing.

When flow is stopped wax crystals form, causing the crude to gel in the pipeline. The wax in crude which is being pumped at temperatures above its pour point will form cohesive lattice structures if it is allowed to cool down to below its pour point whilst stationary. Experiments have shown that restart pressures can be five to ten times higher for a pipeline that was above the pour point and cooled after shut-down than for one that was below its pour point before shut-down.

Gas Pipelines

Several formulae can be used to calculate the flow of gas in a pipeline. These formulas account for the effects of pressure, temperature, pipe diametre, pipe length, specific gravity, pipe roughness and gas deviation.

The Darcy equation can also be used in flow calculations involving gases but it must be done with care and restrictions on its use are recommended. If, for instance, pressure drop in the line is large relative to the inlet pressure, the Darcy equation is not recommended. Because this is often the case and because other restrictions also apply to its use in gas flow calculations, other more practical equations are commonly used for gas flow calculations.

Allowable Operating Pressure

An important pipeline design calculation is the maximum pressure at which a given size, grade and weight of pipe may operate.

Maximum operating pressure determines how much a pipeline may carry. Other factors being fixed and depends on the physical and chemical properties of the pipe steel. Since standard pipe grades, sizes and weights are normally used, the maximum operating pressure can usually be obtained from tables contained in recognised specifications.

Looping

This is the term used when laying a pipeline parallel to an existing line in order to increase the total capacity throughput.

Two-phase Flow

The combined flow of oil and gas in a pipeline presents many design and operational difficulties not present in single phase liquid or vapour flow. Frictional pressure drops are harder to estimate.

Liquid is likely to gather at low points in the pipeline and reduce the pipeline capacity to a point when slugs of liquid are pushed ahead by the gas. The movement of large liquid slugs along the pipeline can cause additional pipeline stresses and the pipeline terminal facilities must be designed to receive such volumes of liquid by provision of large, specially designed vessels or energy absorbing pipework, known as slug-catchers. The type of flow in a pipe is known as its flow regime. We have already come across laminar and turbulent flow regimes. These are single phase flow regimes and which phase will exist can be found by calculating the Reynolds number. Pipelines are seldom horizontal, as they have to follow the undulations of the seabed or the countryside, and often have vertical sections as they rise to join platforms or enter process streams. In view of this, flows regimes can exist which are considerably more complex than those already discussed.

The key difference between single-phase flow and two-phase flow is that it is much more difficult to determine pressure drops for two-phase flow. This is complicated if you consider that a difference in incline of several degrees, never mind 90°; can change entirely the nature of the flow regime.

Undulating terrain will generally not be a problem for single-phase pipelines; however, it can materially affect pressure drop in two-phase pipelines if there are a large number of. rises and falls, which the pipeline must cross.

Some two-phase regimes are caused by liquid condensation or fall-out from the gas due to reducing temperature and pressure along the length of the pipeline. For onshore gas lines liquid knock-outs can be provided at intervals such that liquids can be drained off by blow-down of the line. Well flow lines often work in a two-phase regime, particularly because the well fluids usually contain both oil and gas and there may be no facility at the wellhead (E.g. at sub-sea wells) prior to the fluid reaching the gathering station (or platform).

Despite the problems associated with the prediction of two-phase estimates, more and more pipelines are being designed for such flow systems. For example when hydrocarbon condensate is separated from the gas at offshore platforms, it is invariably spiked back into the gas for transport to the shore in the pipeline.

This is mainly because te economics would not support a separate line for condensate sales. Several empirical flow patterns have been presented that determine vapour/liquid flow as a function of fluid proportions and flow rates. Diagrams of these flow patterns are shown Figure.

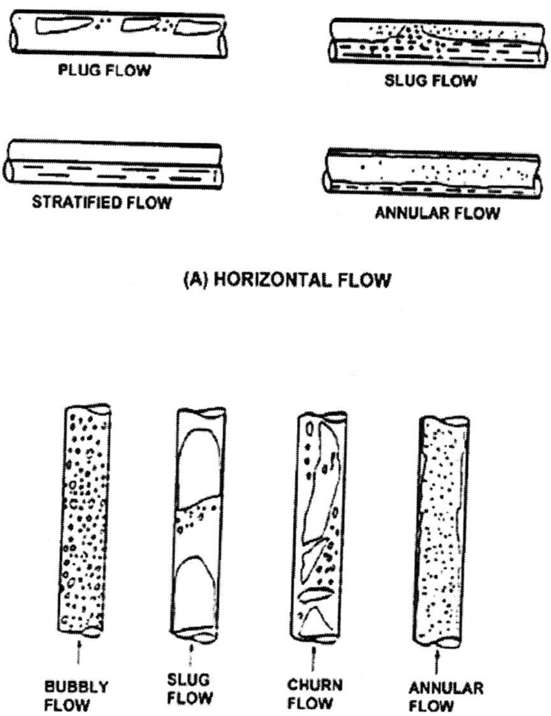

Figure : Pipe line flow patterns

Care should be taken in the interpretation of these diagrams, as the regime boundaries of bubble, slug, annular, mist and wave conditions are strongly affected by pipe inclination. Even very low pipe inclination of one or two degrees can cause considerable movement of the regime boundaries and, in addition, adjustment has been observed due to fluid pressure, pipe diametre and surface tension.

In both vertical and horizontal directions, the avoidance of slug flow is desirable. Slug flow might possibly be avoided by choice of a smaller pipe diametre. This will increase fluid velocities and reduce the pipeline liquid inventory.

Hydraulic Head

Hydraulic head or piezometric head is a specific measurement of water pressure above a geodetic datum. It is usually measured as a water surface elevation, expressed in units of length, at the entrance (or bottom) of a piezometre. In an aquifer, it can be calculated from the depth to water in a piezometric well (a specialised water well), and given information of the piezometre's elevation and screen depth. Hydraulic head can similarly be measured in a column of water using a standpipe piezometre by measuring the height of the water surface in the tube relative to a common datum. The hydraulic head can be used to determine a *hydraulic gradient* between two or more points.

"Head" in Fluid Dynamics

In fluid dynamics, head is a concept that relates the energy in an incompressible fluid to the height of an equivalent static column of that fluid. From Bernoulli's Principle, the total energy at a given point in a fluid is the energy associated with the movement of the fluid, plus energy from pressure in the fluid, plus energy from the height of the fluid relative to an arbitrary datum. Head is expressed in units of height such as metres or feet. The static head of a pump is the maximum height (pressure) it can deliver. The capability of the pump can be read from its Q-H curve (flow vs. height). Head is equal to the fluid's energy per unit weight. Head is useful in specifying centrifugal pumps because their pumping characteristics tend to be independent of the fluid's density. There are four types of head used to calculate the total head in and out of a pump:

- *Velocity head* is due to the bulk motion of a fluid (kinetic energy).
- *Elevation head* is due to the fluid's weight, the gravitational force acting on a column of fluid.

- *Pressure head* is due to the static pressure, the internal molecular motion of a fluid that exerts a force on its container.
- *Resistance head* (or *friction head* or Head Loss) is due to the frictional forces acting against a fluid's motion by the container.

Components of Hydraulic Head

A mass free falling from an elevation $z>0$ (in a vacuum) will reach a speed

$v=\sqrt{2gz}$, when arriving at elevation z=0, or when we rearrange it as a *head*: $h=\dfrac{v^2}{2g}$

where

$g$is the acceleration due to gravity

The term $\dfrac{v^2}{2g}$ is called the *velocity head*, expressed as a length measurement. In a flowing fluid, it represents the energy of the fluid due to its bulk motion.

The total hydraulic head of a fluid is composed of *pressure head* and *elevation head*. The pressure head is the equivalent gauge pressure of a column of water at the base of the piezometre, and the elevation head is the relative potential energy in terms of an elevation. The *head equation*, a simplified form of the Bernoulli Principle for incompressible fluids, can be expressed as:

$$h=\psi+z$$

where

- $h$is the hydraulic head (Length in m or ft), also known as the piezometric head.
- $\psi$is the pressure head, in terms of the elevation difference of the water column relative to the piezometre bottom (Length in m or ft), and
- $z$is the elevation at the piezometre bottom (Length in m or ft)

In an example with a 400 m deep piezometre, with an elevation of 1000 m, and a depth to water of 100 m: z = 600 m, ψ = 300 m, and h = 900 m. The pressure head can be expressed as:

$$\psi=\frac{P}{\gamma}=\frac{P}{\rho g}$$

where

- P is the gauge pressure (Force per unit area, often Pa or psi),
- γ is the unit weight of water (Force per unit volume, typically $N \cdot m^{-3}$ or lbf/ft^3),
- ρ is the density of the water (Mass per unit volume, frequently $kg \cdot m^{-3}$), and g is the gravitational acceleration (velocity change per unit time, often $m \cdot s^{-2}$)

Fresh Water Head

The pressure head is dependent on the density of water, which can vary depending on both the temperature and chemical composition (salinity, in particular). This means that the hydraulic head calculation is dependent on the density of the water within the piezometre. If one or more hydraulic head measurements are to be compared, they need to be standardised, usually to their *fresh water head*, which can be calculated as:

$$h_{fw} = \psi \frac{\rho}{\rho_{fw}} + z$$

where

- h_{fw} is the fresh water head (Length, measured in m or ft), and
- ρ_{fw} is the density of fresh water (Mass per unit volume, typically in $kg \cdot m^{-3}$)

Hydraulic Gradient

The hydraulic gradient is a vector gradient between two or more hydraulic head measurements over the length of the flow path. It is also called the Darcy slope, since it determines the quantity of a Darcy flux, or discharge. A dimensionless hydraulic gradient can be calculated between two piezometres as:

$$i = \frac{dh}{dl} = \frac{h_2 - h_1}{\textbf{length}}$$

where

- i is the hydraulic gradient (dimensionless),
- dh is the difference between two hydraulic heads (Length, usually in m or ft), and
- dl is the flow path length between the two piezometres (Length, usually in m or ft)

The hydraulic gradient can be expressed in vector notation, using the del operator. This requires a hydraulic head field, which can only be practically obtained from a numerical model, such as MODFLOW. In Cartesian coordinates, this can be expressed as:

$$\nabla h = \left(\frac{\partial h}{\partial x}, \frac{\partial h}{\partial y}, \frac{\partial h}{\partial z} \right) = \frac{\partial h}{\partial x}\mathbf{i} + \frac{\partial h}{\partial y}\mathbf{j} + \frac{\partial h}{\partial z}\mathbf{k}$$

This vector describes the direction of the groundwater flow, where negative values indicate flow along the dimension, and zero indicates *no flow*. As with any other example in physics, energy must flow from high to low, which is why the flow is in the negative gradient. This vector can be used in conjunction with Darcy's law and a tensor of hydraulic conductivity to determine the flux of water in three dimensions.

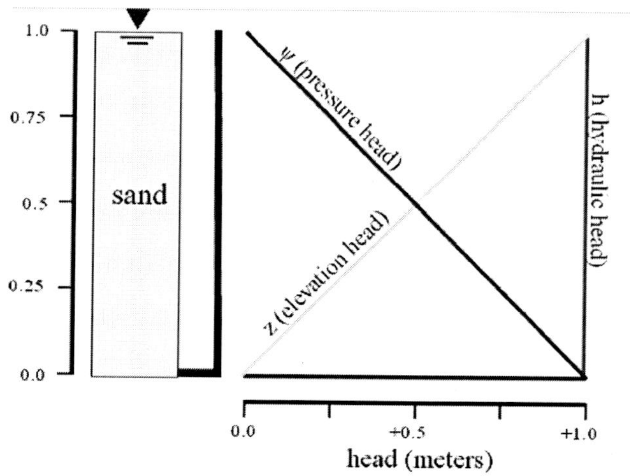

Figure: Relation between heads for a *hydrostatic* case and a *downward flow* case.

Bibliography

Beer, F.P., Johnston, Jr., E.R.: *Mechanics of Materials*, McGraw-Hill, Inc., New York, 1981.

Berry, Mary Clay: *Alaska Pipeline: The Politics of Oil and Native Land Claims*. Bloomington, IN: Indiana University Press, 1975.

Bird, John: *Electrical and Electronic Principles and Technology,* Newnes, UK, 2007.

Brownell, W. E.: *Structural Clay Products,* Springer-Verlag Publisher, N.Y., 1976.

Callister, Jr., William D.: *Materials Science and Engineering – An Introduction*, John Wiley and Sons, UK, 2000.

Credi, A.: *Molecular Devices and Machines: a Journey into the Nano World*, Wiley-VCH, Weinheim, 2003.

David Willoughby: *Plastic Piping Handbook*. McGraw-Hill Professional, Delhi, 2002.

Dickenson, T. Christopher: *Valves, Piping & Pipelines Handbook*, Tulsa, OK: PennWell, 1999.

Flynn, M.J. : *Computer Architecture: Pipelined and Parallel Processor Design*, Narosa, Delhi, 2011.

Fred W.: *Sheet Metal Machine Processes*, Delmar, N.P., 1975.

Fukuyama F.: *Our Posthuman Future: Consequences of the Biotechnology Revolution,* Farrar, Straus and Giroux, NewYork, 2002.

Guy, N.G.: *Pipe Protection*. New York, NY: Elsevier Applied Science, 1987.

Hollands, J.: *Engineering Psychology and Human Performance*, Prentice Hall, Delhi, 1999.

Hosmanek, Max: *Pipeline Construction*. Austin, TX: Petroleum Extension Service, 1984.

Hugh Shercliff and David Cebon: *Materials: Engineering, Science, Processing and Design*, Butterworth-Heinemann, UK, 2007.

James O. Pennock: *Piping Engineering Leadership for Process Plant Projects*, Gulf Professional Publishing, 2001.

John J. McKetta: *Piping Design Handbook*. Marcel Dekker, 1992.

John J.: *Piping Design Handbook*. New York, NY: M. Dekker, 1992.

Johnston, Jr., E.R.: *Mechanics of Materials*, McGraw-Hill, Inc., New York, 1981.

Lamit, Louis Gary: *Pipe Fitting and Piping Handbook*. Englewood Cliffs, NJ: Prentice-Hall, 1984.

Lifschitz, E. M.: *Fluid Mechanics*, Pergamon Press, North Holland, 1987.

McCormick, T. : *Metal Technics*, Brooks Waterloo-Jacaranda Press, Milton, Qld, 1989.

Nayyar, P.E., Mohinder L.: *Piping Handbook,* New York: McGraw-Hill, 2000.

Pearson, Norman: *Pipelines and Farming*. London, ON, Canada: Strathmore House, 1995.

Peckworth, H. F.: *Concrete Pipe Handbook,* American Concrete Pipe Assoc., Chicago, 1959.

Peter J.: *Pipelines and Permafrost: Physical Geography and Development in the Circumpolar North*. New York, NY: Longman, 1979.

Pradeep P. Phule: *The Science & Engineering of Materials*, Thomson-Engineering, NY, 2005.

Pulkrabek, W.W. : *Engineering Fundamentals of the Internal Combustion Engine*, Prentice-Hall, Inc., 1997.

Sahu, G.K. : *Handbook Of Piping Design*, New Age International, Delhi, 2009.

Seeman, N.C.: *Biomolecular Stereodynamics*, Adenine Press, New York, 1981.

Sosnin, H.A.: *Procedures for Pipewelding*. Washington, DC: National Association of Plumbing, 1983.

Tajima, T.: *Computational Plasma Physics with Applications to Fusion and Astrophysics*, Westview Press, US, 2004.

Utkin, V.I.: *Sliding Modes in Control and Optimization*, Springer-Verlag, Berlin, 1992.

Van Vlack, L. H.: *Elements of Materials Science,* Addison-Wesley Publishing Co., Reading, Mass., 1967.

Warring, R.H.: *Handbook of Valves, Piping, and Pipelines*. Houston, TX: Gulf Publishing Co., 1982.

Index

❑❑❑